Susanne Klein-Vogelbach

Functional Kinetics

Observing, Analyzing, and Teaching Human Movement

With 329 Illustrations

Springer-Verlag Berlin Heidelberg New York
London Paris Tokyo Hong Kong

Dr. med. h. c. Susanne Klein-Vogelbach
Felixhäglistraße 12
CH-4103 Bottmingen

Translator:
Gertrud Whitehouse
Yew Tree Cottage
Car Colston Road
Screveton
Nottinghamshire NG13 8JL, UK

Translation of the 4th German edition:

Funktionelle Bewegungslehre

© Springer-Verlag Berlin Heidelberg 1976, 1977, 1984, 1990

ISBN 3-540-15350-0 Springer-Verlag Berlin Heidelberg New York
ISBN 0-387-15350-0 Springer-Verlag New York Berlin Heidelberg

Library of Congress Cataloging-in-Publication Data
Klein-Vogelbach, Susanne, 1909- [Funktionelle Bewegungslehre. English]
Functional kinetics : observing, analyzing, and teaching human movement / Susanne
Klein-Vogelbach ; [translator, Gertrud Whitehouse]. p. cm. Translation of the 4th ed.
of: Funktionelle Bewegungslehre. Includes bibliographical references.
ISBN 0-387-15350-0 (U. S. : alk. paper)
1. Exercise therapy. I. Title. [DNLM: 1. Exercise Therapy. 2. Kinetics.
3. Movement. WB 541 K64f]
RM725.K5413 1990 615.8'2--dc20 DNLM/DLC 89-21745

© Springer-Verlag Berlin Heidelberg 1990
Printed in Germany

Typesetting, printing and binding: Appl, Wemding
2121/3145-543210 - Printed on acid-free paper

Preface to the English Edition

The translation of my book *Funktionelle Bewegungslehre* into English proved an experience unexpectedly rich in learning. When one has no more than a poor command of the language into which one's own work is being translated, one tends – perhaps out of laziness – to assume that one is not competent to judge the result. That was what I thought, anyway.

The English Copy Editing department at Springer-Verlag, in the person of Kersti Wagstaff, taught me better. At her encouragement, armed with dictionaries, I sat down and went through the translation and my German original line by line, making corrections in my stumbling English where the translation did not quite express my intentions. Kersti Wagstaff then spared no efforts in grappling with my ideas so as to be able to rewrite my corrections in clear English. At many places in the manuscript I was told: 'I don't understand what you mean here' – and behold, when I looked again at the German I had to admit that that particular passage could be interpreted in several different ways.

Through this – to me, ideal – process of collaboration, I received invaluable objective criticism of my book, and for this I owe Bernhard Lewerich, who looks after my books at Springer-Verlag, and, especially, Kersti Wagstaff a great debt of thanks.

Bottmingen, August 1989   Susanne Klein-Vogelbach

Preface to the Fourth German Edition

This fourth edition has arisen out of work done on the translation of *Functional Kinetics* into English. Working on that book with Kersti Wagstaff of Springer-Verlag's copy editing department, I was able to go back and improve and clarify many places in the German text.

In addition, the *eight observation criteria of normal gait,* as presented in *Gangschulung zur funktionellen Bewegungslehre* (shortly to be published by Springer-Verlag in the series *Rehabilitation und Prävention,* Vol. 16), have been taken into account:

1. Forward transportation of body segments thorax and head with their frontotransverse axes horizontal and at right angles to the direction of locomotion
2. Step rate
3. Track width
4. Step length
5. Maintenance of the virtual long axis of the body and its vertical alignment
6. Maintenance of the virtual axes of the foot and the alignment of the functional long axis of the foot in the direction of locomotion
7. The walking patterns of body segments legs and pelvis
8. The armswing as *reactio* to the walking patterns of body segments legs and pelvis

My special thanks are due to Bernhard Lewerich of Springer-Verlag, who has met my many requests as an author with great generosity.

Bottmingen, July 1989   Susanne Klein-Vogelbach

Preface to the Third German Edition

The concept of 'functional kinetics' has attracted many supporters and collaborators since its first appearance 7 years ago.

During the teaching of the theory of functional kinetics in training schools and on courses, problems of interpretation arose which were unknown to me, because the concept was my own, and when teaching it myself, I was always able to provide the answer to any queries that came up. I have accepted the queries and problems of my colleagues and students in the spirit of constructive criticism and have appreciated their valuable contribution.

This has led to a complete revision of this book. I have tried to reshape the material in such a way that it is no longer merely an adjunct to my own teaching but an independent presentation of the concept of the observation and analysis of movement on which to base a therapy of movement truly proper to the human being.

This book is intended as an introduction to the theory of functional kinetics. For this reason extensive citation of references has been omitted.

My thanks are due to:
- Springer-Verlag, who made this new edition possible.
- Ortrud Bronner, principal of the School of Physiotherapy at the Kantonspital Basel, who has given a great deal of her time to assist with revision of the manuscript, and Barbara Bartmes, principal of the School of Physiotherapy in Mannheim.
- Katrin Eicke-Wieser for her help in proof-reading and correcting the manuscript, a time-consuming task.
- Adrian and Gudrun Cornford for designing the illustrations for this edition, Verena Sofka for those of the first and second editions, and Ursula Künzle and Ruth Thierstein for the photographs; Andreas Bertram and Jacques Hochstrasser for the presentation of the photographs.
- My colleague Isabella Gloor-Moriconi and all others, too numerous to mention, who have assisted with the preparation of this book.

Basel, November 1983   Susanne Klein-Vogelbach

Preface to the First German Edition

For as long as I can remember, the beauty of movement, human or animal, has fascinated me.

As a teacher of gymnastics I learned to see that healthy and natural movement of a living creature is beautiful, and beautiful movement became my guiding principle in physical education. Confrontation with the reality of the differences in people's gifts for movement, however, has taught me to ask myself why any given sequence of movements should be natural and effortless to some people while others cannot achieve it even after dedicated practice.

Searching for reasons to explain these differences, I was forced to start observing, and began to realize that the constitution, mentality and physical condition of a human being determine whether he is predisposed or indisposed to perform any particular physical activity; suitability (or unsuitability) for any type of movement must, therefore, be predictable. Thus I came to appreciate how relative the concepts 'normal', 'healthy' and 'sick' are. The step to therapy was small and changed my concept of movement training only gradually. My model remained the same; only the road to achieving it became longer, sometimes more arduous, and on occasion impassable.

In this book I have tried to order the experience of a lifetime as a professional teacher of movement and to point towards a possible way of considering movement systematically, of observing and analysing it, and of conveying the findings in such a manner that the patient can learn from them.

We should never forget that human understanding is insufficient to grasp the phenomena of movement in their entirety. But we can observe them and come to recognize their outward form.

My thanks for advice and collaboration to: Georg Klein-Vogelbach, my husband; Gisela Rolf and Irmgard Flückiger; Verena Sofka-Lagutt, graphics; Ortrud Bronner, Katrin Eicke-Wieser, Verena M. Jung, readers; and Heidi Säckinger-Wolf and Anne Schäfer, secretaries.

Basel, February 1976   Susanne Klein-Vogelbach

Note on Pronouns

For clarity and brevity, 'she' has been assigned to the physiotherapist and 'he' to the patient.

Contents

General Introduction

The theory of functional kinetics has evolved from practical experience, developed during the process of treating patients and teaching students. It is important to see and feel what actually happens when the human body is in motion, and to recognize when and why deviations occur. The perception of deviation presupposes that we possess a mental image of normal movement and consequently an ability to recognize deviations from it. It follows that the theory of functional kinetics cannot be derived from pathological symptoms. Pathological symptoms cannot provide the model because they are a deviation from a healthy state, and as it is our aim to pursue a form of physiotherapy proper to the natural human being, it is only reasonable to choose the normal movement patterns of a healthy human being as our model.

Life is Movement
If life is movement, promoting movement is conducive to a better life; it is therefore the physiotherapist's task to stimulate movement.

Learning Goal
We have developed a technique which therapists can learn and which will enable then to improve an individual's vitality, using instruction and manipulation to teach the patient how to move.

We should take note of the following:
- Natural movement is carried out automatically.
- The aim, intention and planning of movement may be volitional.
- In the deliberate performance of goal-orientated movement, we make use of patterns of movement which are automatic.
- We become conscious of the events in movement when we are tired.
- We become conscious of how a movement is performed when we are not, or not yet, competent at it.
- We become conscious of movement which demands unusual exertion.
- Our consciousness is involved when we employ movement to make ourselves understood or as a means of expression.
- Not all people possess the same repertoire of movement. This repertoire is dependent upon the individual's natural disposition, environmental factors and repetition, i. e. practice.
- Every mobile and healthy human being can move effortlessly by his own strength. He can use his hands within the range of activities necessary for self-preservation. He can acquire skills, comprehend his environment through the senses of sight,

1

hearing, smell and touch, and can communicate effortlessly with others who speak the same language.

Note

Even faulty movement may become automatic, but it lacks the quality of naturalness and is not economical. A movement is economical when performance and result are maximal and effort and wear and tear are minimal.

Those who move naturally experience movement as something axiomatic and agreeable. To an observer natural movement is harmonious and therefore beautiful. Faulty movement is perceived as disturbing and unnatural. A deviation from what is normal is always an avoiding action, a form of limping. It is immaterial whether the avoidance mechanism has been triggered by pain, restricted mobility, dysfunction, trauma, psychological disturbance or merely bad habits.

What then is the task of the physiotherapist? What does she need to stimulate mobility in a manner proper to the human being? The physiotherapist must be able to analyse movement, in order to find and define the functional problem. She must be able to effect the required change in movement, whether through manipulation, instructing the patient how to move, or both.

The physiotherapist looks for characteristic features within the complexity of the pattern of movement using her powers of observation and skill in palpating, but no extrinsic aids (e. g. measuring instruments). The inborn talent of all living creatures to recognize whether another of its kind is normal and to distinguish it from what is unhealthy is for the physiotherapist an invaluable and indispensable natural gift. On the basis of this, and with the aid of orderly reasoning, she will learn to develop a functionally appropriate therapy of movement.

The theory of functional kinetics thus presents a technique of the direct observation of movement and of its evaluation for therapy. This approach appears simple; what is complicated is the great subtlety and differentiation of normal movement. The possibilities are unlimited, and for that very reason definitively perfect results can never be achieved.

1 Orientation of the Individual

If a human being cannot orientate himself within his own body, within space, and outwards in relation to his own body, his perception is impaired, and he cannot move normally.

We are concerned here with the orientation of the individual, a factor of equal importance to both therapist and patient. Therapist and patient have similar sensory perception, and when they move, similar information is transmitted to their central nervous systems. However, when the therapist takes on the role of a teacher and has to guide the patient during treatment, the situation is different, and difficulties in communication can arise. If the instructions she gives in the course of manipulation or movement training are expressed from her point of view as an observer, the patient may be unable to carry them out. To be effective, her directions need to be given in language which is meaningful to the patient. Only then can the patient do what is asked.

● *Examples*
Wrong instruction: 'Stretch out your right arm.'
Right instruction: 'Move your right hand as far as possible away from your right shoulder.'
Wrong instruction: 'Raise your left arm.'
Right instruction: 'Move your left hand upwards as far as you can, so that it is higher than your head.'

This chapter will show how an individual orientates himself by information received from his perceptive faculties. If we understand this, we will be able to find the right words at the right time when dealing with a patient.

> **Note**
> The capacity for orientation is partly innate and partly acquired. It develops when a child begins to raise its head, sit up, stand and walk.

1.1 Orientation Within the Body

> **Note**
> Orientation within one's own body is achieved through kinaesthetic perception, particularly deep sensitivity. It is not determined by environmental factors.

3

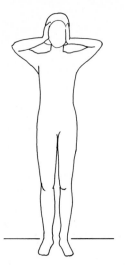 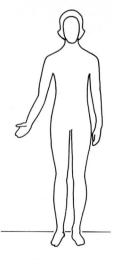 

Fig. 1. Where is your head?

Fig. 2. Where is your right hand?

Fig. 3. Where is your right foot?

Fig. 4. Where does your left hand begin?

Fig. 5. Where does your left arm begin?

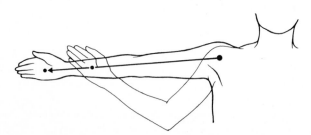

Fig. 6. Move your right wrist away from your right shoulder

Through *static kinaesthetic sensitivity* we perceive the position of parts of the body and the distance between any points on the body.

Through *dynamic kinaesthetic sensitivity* we perceive the direction of movements and alterations in the distance between points on the body.

● *Examples*

Perception of position: We know, without a mirror and without looking, where our head, right hand, right foot are (Figs. 1–3), and we know this whether we are moving or standing still. Without thinking or looking we can grasp any part of our body with our hands, as far as our mobility permits.

We can also indicate demarcation lines between different parts of our body (Figs. 4, 5). We know the relative position of our joints and limbs, and we can feel the tension in our muscles by touching them. We know with our eyes shut whether our hand is clenched in a fist, or whether our hands are clasped and lying in our lap.

Perception of distance: In general, we can carry out the following instructions: 'Indicate with your hands the distance between your right and left shoulder joints'; 'About how far away is your navel from your right earlobe when you are lying on your back?' 'Move your feet apart so that they are in line with your shoulders.'

Perception of direction: Here, too, it is easy to follow instructions: 'Move your hands towards your feet'; 'Move your right knee towards your head'; 'Place the palm of your right hand on your abdomen'; 'Move your left hand outwards from your head and backwards'; 'Move your hands outwards from your abdomen and sideways.'

Perception of alteration in distance: We can also follow the instructions: 'Move your right wrist away from your right shoulder' (Fig. 6); 'Move your left foot as close as possible towards your hip' (Fig. 7); 'Move your right hip as close as possible towards your right heel' (Fig. 8).

Orientation within one's own body provides a number of concepts suitable for instructions requiring specific movement, because of their immediate appeal to the patient's awareness. These are the 'right words'. They are indispensable to *patient language* (see p. 14).

Indicating place	*Indicating direction*
On the abdomen	'Abdomen-wards' (in the direction in which the abdomen is *facing*)
On the back	'Back-wards'
On the head	'Head-wards'
On the foot	'Foot-wards'
On the right side	To the right
On the left side	To the left
In the middle	Towards the middle
	Away from the middle

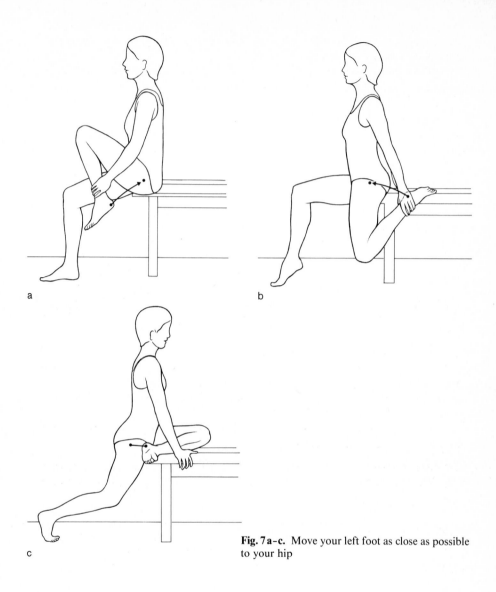

Fig. 7a-c. Move your left foot as close as possible to your hip

1.2 Orientation in Space

> **Note**
> The body's orientation in space is determined by the force of gravity. From this is derived the concept of *up* and *down*.

Because of gravity, the individual experiences his own weight at his points of contact with the environment. There are several ways in which he may perceive his weight through these points of contact:

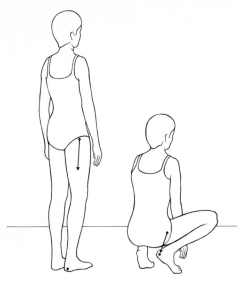

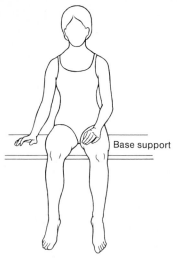

Fig. 9. Perception of pressure at the points of contact between the body and a base support

Fig. 8. Move your right hip as close as possible to your heel

- As pressure at the points of contact with a base support (Fig. 9)
- As pressure with a greater or lesser tendency to slide at the points of contact when the body is leaning against a supportive device (Fig. 10)
- As pull with a greater or lesser tendency to slide at the body's points of contact when hanging from a suspension device (Fig. 11).

Orientation in space likewise provides terms suitable for giving the patient instructions because they appeal immediately to the patient's perception. They are the 'right words' for dealing with a patient.

Indicating place	*Indicating direction*
Up	Upwards (against the force of gravity)
Down	Downwards (with the force of gravity)

The way an instruction is carried out may vary depending upon the starting position. If, for instance, we ask an ordinary person to raise his arms, they will describe an arc twice as large if he raises them while standing as they will if he raises them while lying down (Figs. 12, 13).

The direction of the effect of gravity is constant, whereas the body can change position in space. The human body is constantly operated on by the force of gravity, regardless of its position and whether it is at rest or in motion. When we stand or walk our head is uppermost and 'faces' upwards, our feet are down and 'face' downwards (Fig. 14). If we stand on our head, our head is down and faces downwards while the feet are up, facing upwards (Fig. 15). When we lie on our back, our back is underneath and faces downwards, and our abdomen is uppermost, facing upwards (Fig. 16). When we per-

Fig. 10. Perception of pressure with a tendency to slide at the points of contact between the body and a supportive device

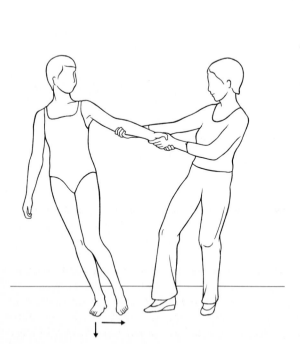

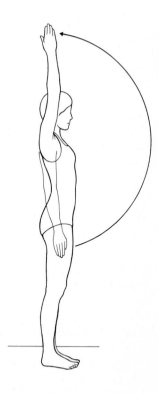

Fig. 11. Perception of pull with a tendency to slide at the points of contact between the body and a suspension device

Fig. 12. Raising the arms when standing

8

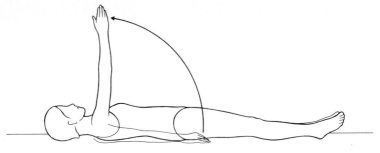

Fig. 13. Raising the arms when lying down

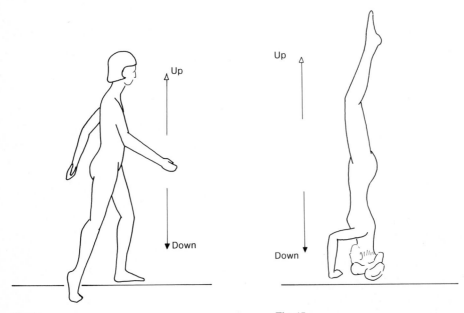

Fig. 14

Fig. 15

form an exercise like the horizontal balance position (Fig. 17), our abdomen is underneath, facing downwards, and our back is on top, facing upwards.

Instructions Which Refer to the Patient's Perception of Contact with the Environment

Change of pressure at the body's points of contact with a base support: 'You are standing upright. The soles of your feet are in contact with the floor in line with your trunk. Push the sole of your right foot against the floor and feel how the sole of your left foot loses contact with the floor.' The result is the stance shown in Fig. 18, in which the whole weight of the body is supported on one leg.

9

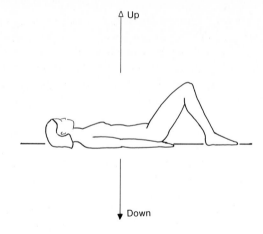

Fig. 16

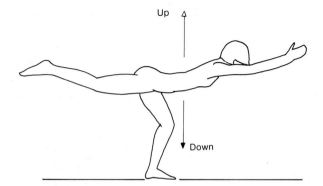

Fig. 17

Fig. 18. Standing on one leg, arrived at by increasing pressure under the sole of the right foot

Change of pressure with a tendency to slide at the body's points of contact with a supportive device: 'Stand upright facing the wall and two feet away from it. Now feel the pressure of the balls of your feet upon the floor increasing while your heels lose contact with the floor. To avoid falling, support yourself by placing the palms of your hands against the wall at the level of your rib-cage. Your body should remain straight, your feet and hands must not be allowed to slide.' (Fig. 19.)

Change of pressure with pull and a tendency to slide at the body's points of contact with a base support and suspension device: 'Open the door. Stand close to its edge so that your abdomen, thorax and forehead are touching it. Hold on to the door handles with both hands. Now, let your hands slide up the door and fold them over the upper edge. Take a short step forwards (half a footlength) with each foot and let your abdomen slide downwards along the door until you feel that you are suspended by your hands. The pressure on the soles of your feet is reduced, but you must not let your feet leave the floor.' (Fig. 20.)

Note
The purpose of reactive muscle activity, invoked by the force of gravity, is to maintain posture and in so doing to prevent the body, which is inherently mobile, from falling.

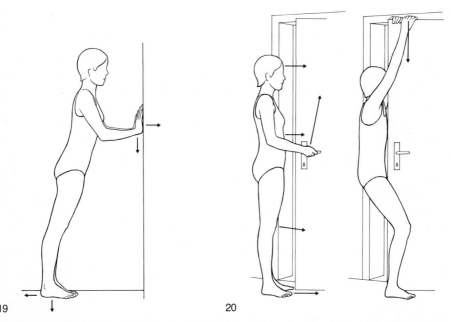

19 20

Fig. 19. Leaning: there is a tendency for the hands to slide against the wall and the feet to slide along the ground

Fig. 20. Hanging with a tendency to slide from a door

Muscle activity within the range of accustomed intensity is not perceived by the individual, but one can feel it on touching one's own body. With a little practice it is possible to increase or reduce the intensity of the activity of the muscle that has been touched, or almost totally inhibit any activity of it, as long as this muscle has received no stimulus to act against the effect of gravity (see p. 247).

1.3 Orientation Outwards from the Body

The characteristic features of human posture are that the long axis of the body is almost vertically aligned, the head is on top, and the eyes are positioned horizontally. This is the normal posture for normal activities and movement, from which people orientate themselves outwards from their body with the aid of visual, proprioceptive, auditory and other modes of sensory perception.

Note
Orientation outwards from the body is determined by the individual's field of vision when upright. It relates to the environment and divides space into segments horizontally. The definition of the direction of the effect of gravity as a spatial constant, independent of a person's posture, is followed by this horizontal segmentation of space according to the direction in which a person (when sitting or standing upright) is looking, conceived in terms of front and back, left and right.

Differentiation of horizontal space is essential, for example, to locomotion, for locomotion involves continuous changes of the body's points of contact with the base support. Since locomotion in space implies a continuous change of location, it is obviously most economical when its direction is forwards in a horizontal straight line.
Orientation outwards from the body also gives us a number of terms which lend themselves well to movement instructions because they appeal directly to the patient's perceptions.

Indicating place	*Indicating direction*
In front	Forwards
Behind	Backwards
Right	(Towards the) right
Left	(Towards the) left

In front is our field of activity: our eyes look forwards, we walk forwards, and we use our hands mostly in front of our body (Fig. 21).

Behind is our field of nonactivity. We have no vision to the rear. In locomotion, it is the place from which we have come (Fig. 22).

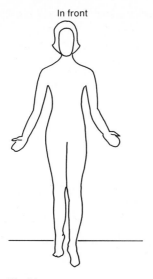

In front

Fig. 21

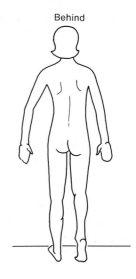

Behind

Fig. 22

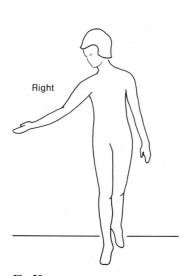

Right

Fig. 23

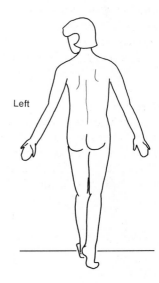

Left

Fig. 24

Right is the dominant side of the field of vision for right-handers, because it is the area in which they have greatest skill (Fig. 23).

Left is the secondary side; the extremities on the left-hand side are symmetrical partners to the extremities on the right (Fig. 24).

13

1.4 Orientation of the Individual and 'Patient Language'

When a person is supine, orientation in space and orientation within the body predominate. Patients who have been confined to bed for a long time have no opportunity to orientate themselves outwards from the body and lose their ability to do so, causing *in front* to become identical with *up* (Fig. 25). A healthy and active human being, on the other hand, even when lying down feels *in front* to be where it would be if he were standing up (Fig. 26).

Note
The orientations of the individual provide the physiotherapist with the essential vocabulary of 'patient language'.

As long as the patient is supine, it is advisable to speak in terms relating to orientation within the body when giving instructions.

● *Example*
The patient sits upright on a stool. The soles of his feet are in contact with the ground and are the width of the pelvis apart. The knees and toes point forwards. The right hand is on the right thigh and the left on the left thigh. The right hand now moves slowly forwards and upwards, away from the right shoulder, which does not change its position in space. The movement is concluded when the hand, with fingers pointing forwards and palm facing downwards, reaches a position in front of the right shoulder, a little to the right. This pattern of movement is repeated several times until it has been impressed in the memory. Then follows an exercise practising movements of the left

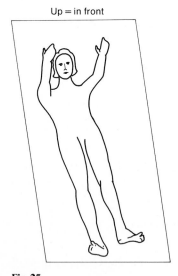

Up = in front

Fig. 25

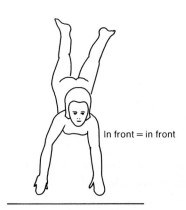

In front = in front

Fig. 26

hand. The fingertips of the left hand touch the middle of the lower abdomen, the fingertips placed one above the other with the little finger at the bottom. The fingers feel their way up the middle of the rib-cage and the profile of the face until the little finger loses contact with the forehead. The hand keeps on moving further upwards and backwards until it is above the head with the palm of the hand facing the crown of the head. The fingers should try to follow the contours of the head. This movement also needs to be repeated until the 'movement image' has been impressed in the memory.

From the same starting position, the movements for left and right should then be performed together, beginning and ending simultaneously. Only then should the speed be increased.

As a movement instruction, the entire movement sequence needs to be translated into 'patient language'. The apparently excessive complication of detail leads the patient immediately and perfectly to perform precisely differentiated movements which will quickly be learnt through repetition. Once a person's perception is sensitized, he can execute quite unconventional movements in response to appropriate verbal instructions.

1.5 Summary

Orientation within the body enables a person to perceive distances between points on his own body through static kinaesthetic sensitivity, and changes in distance between such points through dynamic kinaesthetic sensitivity.

In orientation in space, the individual experiences his body weight as pressure acting upon a base support, as pressure with a tendency to slide when leaning against a supportive device, and as pull with a tendency to slide when hanging from a suspension device.

Orientation outwards from the body leads to perception of movement in space.

2 Orientation of the Physiotherapist

An individual's innate and acquired sense of orientation normally manifests itself as a subliminal awareness of his own body and its movements. This is as true of the physiotherapist as it is of the patient. Since it is the therapist's aim to promote natural and efficient movement, she should make use of the patient's subconscious feeling for movement and his body, guiding him through manipulation ('handling') or instruction by words and gestures as her own insight directs.

These skills can be attained by the physiotherapist in the following ways:

- By improving her own physical fitness and developing her manipulative expertise. This can be achieved through sport, physical exercise and manual activities such as massage, etc. However, in order to be able to experience her own body movements and – by sensitive use of her hands – her environment and the patient, she needs, in addition to physical fitness training (which is often coloured by a spirit of competition), to foster a more receptive attitude. Moving to music, with or without a partner, can be of considerable help in training subtlety of perception.
- Through extensive professional knowledge of the structure and function of the body in health and sickness. The basic concept of treatment in any specific case should emerge from the difference between normal and abnormal movement patterns in the particular patient in question, and thus the framework of a functional treatment plan comes into being.

Note

The *orientation of the physiotherapist* can be defined as profound specialist knowledge of the mechanism and function of the human movement apparatus.

We make use of generally accepted terms from mathematics, physics, anatomy and physiology, and use schemata and observation grids in order to establish rules which enable us to make approximate comparative statements about posture and movement.

To have a special nomenclature for the theory of functional kinetics facilitates clear understanding among physiotherapists. It denotes phenomena of recognized importance in the theory of functional kinetics. It also creates demarcations vis-à-vis established designations for similar but not analogous concepts (e.g. joint as 'switch point of movement').

2.1 Homunculus: The Man in a Cube

We call the person in the cube 'homunculus'. This name is intended to emphasize that we are dealing with a schematization of the human figure and that our statements are approximate (Fig. 27).

Note
1. The cube represents a three-dimensional system, its height being that of the homunculus standing upright.
2. The planes of the cube are transferred to the homunculus (see p. 30). In this way the cube serves as an aid, a grid which facilitates the physiotherapist's orientation in relation to her patient. If there is no special indication otherwise the homunculus is standing upright with its joints at zero (neutral-0 method; see Debrunner 1971).

2.2 Planes – Lines – Points

2.2.1 Transverse Planes

The plane on which the homunculus stands is identical with the face on which the cube rests. In terms of orientation in space, it is the lower tangential plane of the homunculus. The opposite face of the cube touches the homunculus at the vertex and in terms of spatial orientation forms the upper tangential plane of the homunculus.

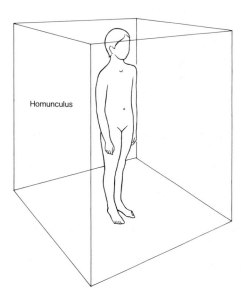

Homunculus

Fig. 27. Homunculus: the man in a cube

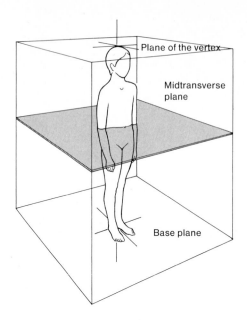

Fig. 28. Transverse planes

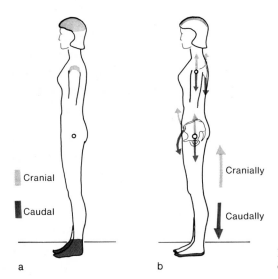

Cranial

Caudal

Cranially

Caudally

a

b

Fig. 29. a Cranial/caudal; **b** cranially/caudally

Any number of parallel planes can be positioned between the upper and lower tangential plane and each one divides the homunculus into a cranial part (pertaining to the head) and a caudal part (pertaining to the tail or feet). All these planes are *transverse planes*. They relate to the body and not to space (Fig. 28).

Note

The transverse plane tangential to the soles of the feet is called the *base plane*.
The transverse plane through the functional centrepoint of the body (see p. 24) is called the *midtransverse plane*.
The transverse plane tangential to the vertex is called the *plane of the vertex*.
When a human being is upright, his transverse planes are *horizontal*. When he lies on his side, on his back or his stomach, his transverse planes are *vertical*.

We have stated that the homunculus is divided into a cranial and a caudal part by every transverse plane. Cranial and caudal are terms in therapist language. In patient language we should rather say, 'belonging to the head/feet'.

Indicating place (positional)

Cranial (on the head or part containing the head)
Caudal (on the feet or part containing the feet)

Indicating direction (directional)

Cranially (towards the head or in the direction in which the head points)
Caudally (towards the feet or in the direction in which the feet point)

The designations 'cranial/caudal' and 'cranially/caudally' are used by the therapist to define precisely the position of parts of the body and alterations in the position of levers at the joints (Fig. 29).

● *Example*
The vertex occupies a cranial position on the body (positional). However, if a patient sits upright on a chair and moves his hand from above his head to rest on the crown of his head, his hand has moved caudally (directional).

2.2.2 Frontal Planes

The homunculus stands inside the cube. One face of the cube is in his field of vision. In terms of orientation outwards from his own body, this face is *in front* of him; the opposite face of the cube is *behind* him.
Any number of parallel planes can be positioned between the front and back sides of the cube. Where they cut the homunculus, they divide him into a *ventral* part (belonging to the abdomen) and a *dorsal* part (belonging to the back). All these planes are *frontal planes* Fig. 30. They relate to the body and not to space.

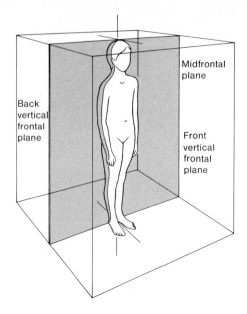

Fig. 30. Frontal planes

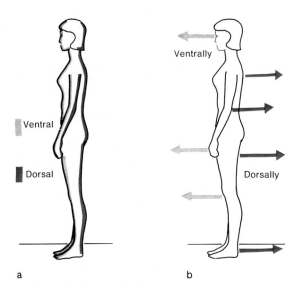

a   b

Fig. 31. a Ventral/dorsal; b ventrally/dorsally

20

Note

The frontal plane through the functional centrepoint of the body (see p. 24) is called the *midfrontal plane*.

The two vertical frontal planes which touch the homunculus ventrally and dorsally are not included among the normal planes of orientation; their points of contact are dependent upon the patient's constitution and therefore have no clear significance in a normative frame of reference. They are, however, very helpful in judging deviation in the postural statics (see Sect. 6.4) when assessing a patient's functional status (see p. 247).

When the patient is upright or on his side, his frontal planes are *vertical;* when he is prone or supine, his frontal planes are *horizontal.*

We have stated that each frontal plane divides the homunculus into a ventral and a dorsal part. 'Ventral' and 'dorsal' are terms in therapist language. In patient language (corresponding to the way the individual orientates himself) we should say, 'belonging to the abdomen/back'.

Indicating place (positional)	*Indicating direction (directional)*
Ventral (on the abdomen or the abdominal part)	Ventrally ('abdomen-wards', in the direction in which the abdomen faces)
Dorsal (on the back or the back part)	Dorsally ('back-wards', in the direction in which the back faces)

The designations 'ventral/dorsal' and 'ventrally/dorsally' are used by the therapist to define precisely the position of parts of the body and alterations in the positions of levers at the joints (Fig. 31).

● *Example*

A gall-bladder operation leaves a ventral scar on a patient's right upper abdomen (positional) but when he sits upright at a table and takes his right hand off the table to place it on the scar, his hand has moved dorsally (directional).

2.2.3 Sagittal Planes

The homunculus stands inside the cube. In terms of orientation outwards from the body, one face of the cube is on the left and one on the right of the homunculus.

Any number of parallel planes can be positioned between the left and right faces of the cube. Where they cut the homunculus, they divide him into a *right lateral* and a *left lateral* part. All these planes are *sagittal planes* (Fig. 32). They relate to the body and not to space.

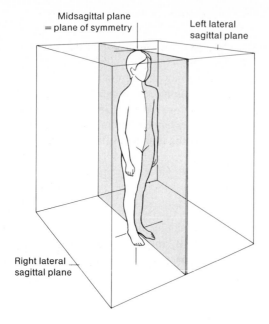

Fig. 32. Sagittal planes

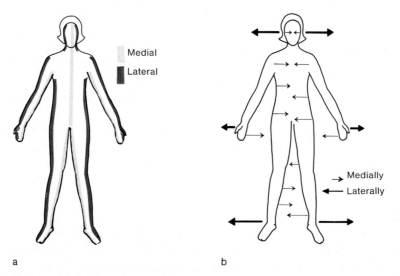

a   b

Fig. 33. a Medial/lateral; **b** medially/laterally

22

We have stated that the homunculus is divided into a right lateral and a left lateral part by every sagittal plane and that division by the plane of symmetry results in symmetrical right and left parts. In patient language we use the terms 'on your right side/on your left side/in the middle'.

Indicating place (positional)	*Indicating direction (directional)*
Lateral (at the side)	Laterally (outwards, sideways)
Right lateral (on the right side)	Laterally towards the right (sideways towards the right, in the direction in which the right side faces)
Left lateral (on the left side)	Laterally towards the left (sideways towards the left, in the direction in which the left side faces)
Medial (inside, in the middle)	Medially (inwards, towards the middle)

The designations 'right lateral', 'left lateral' and 'medial', and 'laterally towards the right', 'laterally towards the left' and 'medially', are used by the therapist to define precisely the positions of parts of the body and alterations in the positions of levers at the joints (Fig. 33).

● *Example*
A patient sits upright on a stool. When the long axes of the thighs point ventrally, the medial sides of the thighs face each other. If the knees are moved apart, the medial borders of the feet lose their contact with the floor and the patient's feet rest on their right and left lateral border respectively. During this process the right knee has moved right laterally/caudally and slightly dorsally; the left knee has moved left laterally/caudally and slightly dorsally.

We speak of a left and a right arm, a left and a right leg, and we differentiate on the arms and legs between a lateral and a medial side; the pelvis, thorax and head all have a left and a right side. The plane of symmetry lies medially.

2.2.4 Body Diagonals and the Body's Functional Centrepoint

A *body diagonal* is a line which connects the centre of a hip joint with the centre of the shoulder joint on the opposite side. The point of intersection of the two diagonals of the body is the body's *functional centrepoint* (Fig. 34).

Note

The midtransverse plane, the midfrontal plane and the plane of symmetry pass through the functional centrepoint of the body. These three planes, therefore, intersect at the body's functional centrepoint. The body diagonals and the centres of hip and shoulder joints are roughly in the midfrontal plane.

Since the centres of the hip joints are symmetrical, they are in the same transverse plane; the same is true of the midpoints of the shoulder joints. The lines connecting the centres of the hip and the shoulder joints respectively are lines of intersection of a frontal and a transverse plane; they are *frontotransverse* axes.

Note

To locate the body's centrepoint, the joints of the shoulder girdle and those of the vertebral column must be in the neutral position (Fig. 35). Movement at the shoulder girdle and in the spinal column alters the position of the body's functional centrepoint (Fig. 36). The functional centrepoint of the body is not identical with the body's centre of gravity. Its position is dependent upon the three values below:

1. The distance between the centres of the shoulder joints.
2. The distance between the centres of the hip joints.
3. The distance between the lines connecting the centres of the shoulder and hip joints respectively. This distance corresponds to the combined height of the pelvis and thorax.

Definition: The 'angle of the body diagonals' is the acute (cranial or caudal) angle formed by the intersection of the diagonals.

Note

The angle between the body diagonals varies but is always less than 90°. The vertex of the angle is identical with the functional centrepoint of the body.

The size of the angle is determined by the distance between the centres of the shoulder and hip joints and the combined height of the pelvis and thorax, i.e. the distance from the symphysis pubis to the suprasternal notch (Fig. 37). Consequently, the size of the angle is also dependent upon individual constitution and postural statics. It is also determined by the position of the shoulder joints in relation to the trunk.

The distance between the centres of the shoulder joints is always greater than the distance between the centres of the hip joints.

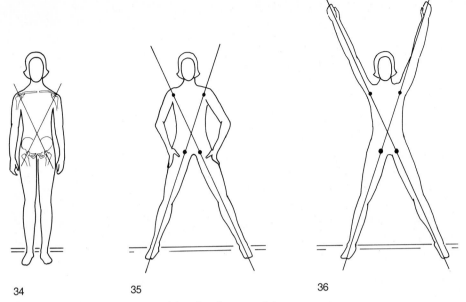

34   35   36

Fig. 34. The body's diagonals and functional centrepoint

Fig. 35. The body's centrepoint with the shoulder girdle in the neutral position

Fig. 36. The body's centrepoint with the shoulder girdle raised

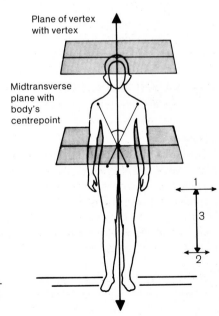

Plane of vertex
with vertex

Midtransverse
plane with
body's
centrepoint

1

3

2

Fig. 37. Influence of the position of *1* the shoulders, *2* the hips and *3* the combined height of pelvis and thorax on the position of the functional centrepoint of the body

- *Examples*

Leg length has no influence on the position of the functional centrepoint of the body but does influence its distance from the ground when a person is standing upright. The functional centrepoint lies most cranially when the shoulders are narrow, the pelvis is broad and high, and the thorax is long (Fig. 38); it lies most caudally when the shoulders are broad, the pelvis is narrow and low, and the thorax is short (Fig. 39). Where the pelvis and thorax are of equal height, the functional centrepoint of a person with broad shoulders and a narrow pelvis (Fig. 40) is lower than that of a person with narrow shoulders and a broad pelvis (Fig. 41).

The size of the angle and the position of the body diagonals are very important in physiotherapy. The study of anatomy teaches us that muscles are optimally stretched and shortened through diagonal movement, since most of them are arranged obliquely. Such diagonal movements can be observed in golf, tennis, shot-putting and throwing the discus and javelin. The diagonal movement patterns of PNF techniques (PNF: proprioceptive neuromuscular facilitation; see Knott 1969) correspond to this given anatomical fact and are for that reason an excellent means of training the musculature.

2.2.5 Vertex and Long Axis of the Body

Definition: The *vertex* of the human being is the point of intersection between the midfrontal plane, the plane of the vertex, and the plane of symmetry.

The *long axis of the body* of the human being is the line of intersection between the midfrontal plane and the plane of symmetry. It passes through the centrepoint of the body and through the vertex (Figs. 37, 42).

The long axis of the body is a *virtual axis*. Virtual body axes are axes which exist only in the imagination but which, within the inherently mobile system of the body, are brought into being by maintenance of a particular posture. Thus the long axis of the body is formed by a particular arrangement of the pelvis, thorax and head, or the long axis of the foot is that (imaginary) one about which the foot arcs when twisting from resting on the medial to resting on the lateral border. By contrast, for instance, the long axis of the thigh is a real and unchangeable axis.

Note

The virtual long axis of the body is an important line with regard to orientation. As the line of intersection of the midfrontal plane and the plane of symmetry, it is a *frontosagittal* axis. When a person is upright it is vertical and its position is closely related to that of the spinal column.

We can use the long axis of the body as a characteristic of normal posture, assuming that the pelvis, thorax and head should be aligned in it.

- *Examples*

We can say of the homunculus: pelvis, thorax and head are aligned in the vertical long axis of the body. The arms hang loosely at the sides, their long axes parallel to the long

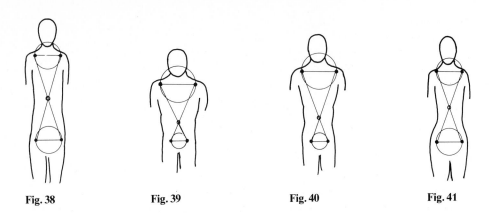

Fig. 38   Fig. 39   Fig. 40   Fig. 41

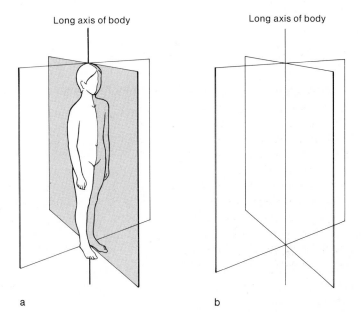

Fig. 42. a The long axis of the body; **b** orientation diagram

axis of the body, and the long axes of the legs are parallel to the caudal projection of the long axis of the body.

The patient sits on a stool. Pelvis, thorax and head are aligned in the long axis of the body which inclines slightly forwards. The long axes of the thighs are horizontal and the knees are apart, the distance between them a little more than the width of the body. The soles of the feet are in contact with the ground, the heels are under the knees, and the toes, like the knees, point forwards and outwards a little to the right and left.

2.2.6 Diameter of the Thorax

Definition: The *frontotransverse diameter of the thorax* is the line of intersection of the midfrontal and transverse planes at the level of the seventh thoracic vertebra (Fig. 43).

The *sagittotransverse diameter of the thorax* is the line of intersection of the transverse plane and the plane of symmetry at the level of the seventh thoracic vertebra (Fig. 44).

Note

The two diameters of the thorax are important defined lines of orientation. We can define any number of body axes as lines of intersection of the body's planes. With their help it is possible precisely to indicate alterations occurring in the body's position in space and/or in the statics of posture, or we can bring about such alterations for therapeutic purposes.

● *Examples*

The line connecting the right and left anterior iliac spines is frontotransverse. When the body is upright in the neutral position, this line is horizontal, at right angles to the body's long axis, and parallel to the frontotransverse diameter of the thorax.

The patient lies on his back. The long axes of the arms are aligned sagittotransversely (Fig. 45).

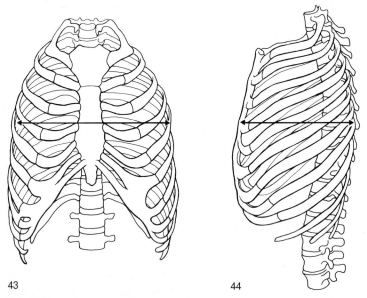

43   44

Fig. 43. Frontotransverse diameter of thorax

Fig. 44. Sagittotransverse diameter of thorax

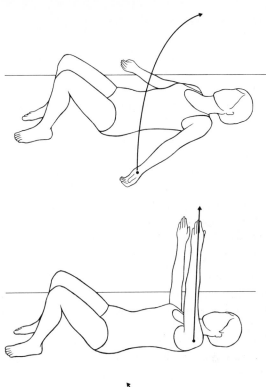

Fig. 45. Sagittotransverse position of the long axis of the arm when the patient is supine

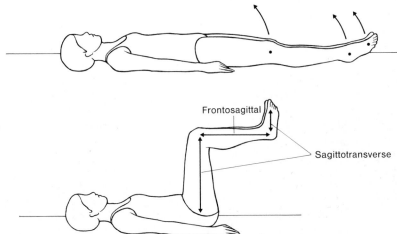

Fig. 46. Patient supine. The long axes of the thighs are sagittotransverse, the long axes of the lower legs frontosagittal, the long axes of the feet sagittotransverse

The patient lies on his back. We place the long axes of the thighs so that they are sagittotransverse, the long axes of the lower legs so that they are frontosagittal and parallel to the midfrontal plane, and the long axes of the feet so that they are sagittotransverse (Fig. 46).

2.2.7 Summary

As an aid to learning, we will review our schemata and observation grids with their points, lines and planes.

Points: The functional centrepoint of the body and the vertex have been defined as points of intersection of three body planes. In Chap. 1 we learned that we become aware of distances between points on the body through static kinaesthetic sensitivity and perceive alterations in distance between such points through dynamic kinaesthetic sensitivity.

Note
A point has no extent and no particular direction (Fig. 47). A point on the body can therefore be guided in an infinite number of directions. The physiotherapist may observe the movement of points or may instruct the patient to bring such movement about.

Straight lines, axes: Axes have been defined as lines of intersection of two planes. The lines of intersection of the body's planes were designated as frontotransverse, fronto-sagittal and sagittotransverse axes. These terms were then used to define the long axis of the body, the diameters of the thorax, and the lines connecting the centres of the shoulder and hip joints; they also serve to describe the position of the long axes of the thighs, the lower legs and the arms.

Note
An axis of the body is a line of intersection of two body planes, or it can be defined as the line connecting two points; definition as the line connecting two points is necessary when the axis is not parallel to one of the co-ordinates of the observation grid. An axis has *one dimension* and extends in *two directions* (Fig. 48). The physio-therapist can observe the position of an axis in relation to space and the other parts of the body; she can observe changes in its position or instruct the patient to bring such changes about.

Planes: To arrive at body planes we considered the sides of a cube in relation to a human being. They alter their position in space with the human being and are defined by the position of the joints in the neutral position. The planes are *transverse, frontal* and *sagittal.* They can intersect the body at any point.

Note
A body plane has *two dimensions* and *four directions* (Fig. 49). It is named according to its position within the body. The physiotherapist can observe the position of the body's planes in relation to space and the position of planes of individual body seg-ments in relation to each other, or she can give instructions to the patient in terms of planes.

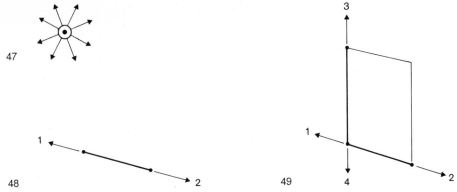

Fig. 47. A point: no dimension and no direction (potential for an infinite number of directions)

Fig. 48. An axis: 2 points, 1 dimension, 2 directions

Fig. 49. A plane: 3 points, 2 dimensions, 4 directions

In the following description of body planes the terms in parentheses are only applicable when the long axis of the body is vertical (see p. 26).

Transverse planes (Fig. 50 a)
Dimensions: frontotransverse and sagittotransverse
Directions: towards the right and the left laterally; ventrally and dorsally (forwards and backwards)
Division: into cranial and caudal (upper and lower)
The frontosagittal axes are at right angles to the transverse planes (vertical).

Frontal planes (Fig. 50 b)
Dimensions: frontotransverse and frontosagittal
Directions: towards the right and the left laterally; cranially and caudally (upwards and downwards)
Division: into ventral and dorsal (front and back)
The sagittotransverse axes are at right angles to the frontal planes (horizontal).

Sagittal planes (Fig. 50 c)
Dimensions: frontosagittal and sagittotransverse
Directions: cranially and caudally (upwards and downwards), ventrally and dorsally (forwards and backwards)
Division: into right and left lateral, and into lateral and medial
The frontotransverse axes are at right angles to the sagittal planes (horizontal).

Cube: Using the frontotransverse, sagittotransverse and frontosagittal lines of intersection, we can project a system of three-dimensional co-ordinates onto any chosen point in the body.

31

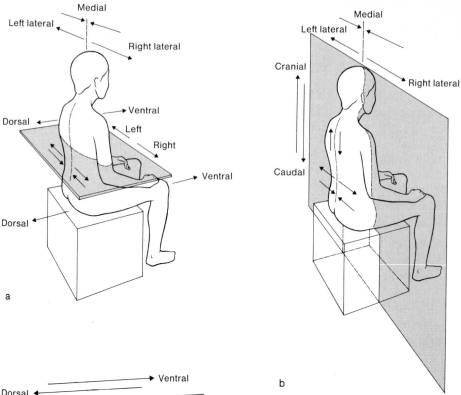

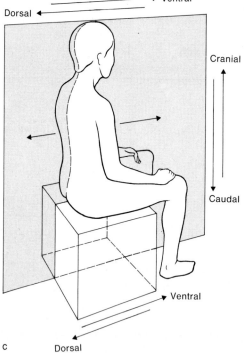

Fig. 50. a Transverse planes, their dimensions and directions. **b** Frontal planes, their dimensions and directions. **c** Sagittal planes, their dimensions and directions

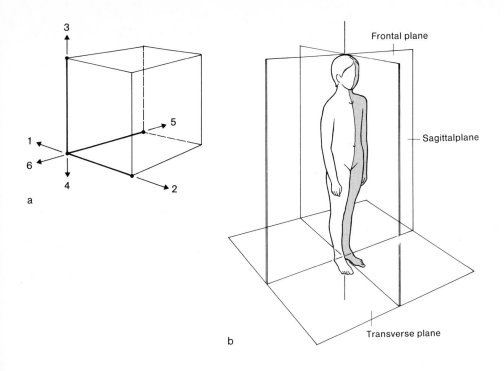

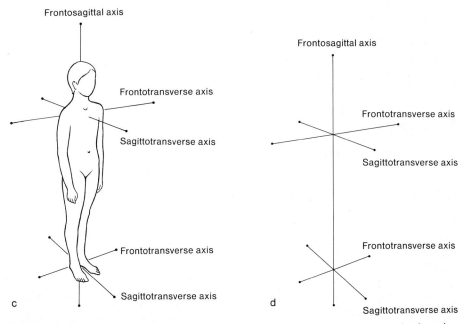

Fig. 51. a The cube: 4 points, 3 dimensions, 6 directions. **b** Planes. **c,d** Axes and orientation schema. **e** Directions

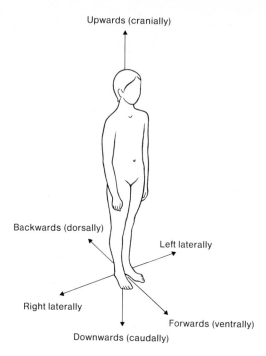

Upwards (cranially)

Backwards (dorsally)

Left laterally

Right laterally

Forwards (ventrally)

Downwards (caudally)

Fig. 51e. See p. 33 for legend

Note

By projecting a three-dimensional system upon the human body, body movements can be reduced to a schema with three dimensions and six directions (Fig. 51).

2.3 Proximal – Distal

Definition: The terms 'proximal' and 'distal' are used with reference to the functional centrepoint of the body. 'Proximal' means 'closer to the centrepoint of the body'; 'distal' means 'further away from the centrepoint of the body'.

Note

'Proximal' and 'distal' are twin concepts which can only be understood in relation to each other. The most proximal zone surrounds the functional centrepoint of the body and is situated in the cranial part of the lumbar spine, roughly in the area of the navel. All other areas of the body are distal to it.

If we follow the links or joints of the skeleton, starting from the cranial part of the lumbar spine, we can trace five routes out from proximal to distal:

1. Functional centrepoint – thoracic spine – cervical spine – head
2. Functional centrepoint – thoracic spine – right via the sternoclavicular joint – acromioclavicular joint – shoulder joint – elbow joint – wrist joint – finger joints
3. Functional centrepoint – thoracic spine – left via the sternoclavicular joint – acromioclavicular joint – shoulder joint – elbow joint – wrist joint – finger joints
4. Functional centrepoint – lower lumbar spine – right via the iliosacral joint – hip joint – knee joint – ankle joints – metatarsal joints – interphalangeal joints of the foot
5. Functional centrepoint – lower lumbar spine – left via the iliosacral joint – hip joint – knee joint – ankle joints – metatarsal joints – interphalangeal joints of the foot.

● *Examples*
The pelvis is caudal or distal to the lumbar spine.
The knees are caudal or distal to the pelvis.
The forearm is proximal to the hand.
The upper arm is proximal to both the forearm and the hand.
The thoracic spine is proximal to both the shoulder girdle and the hand.
The thoracic spine is proximal to both the cervical spine and the head.

Body points sited such that the functional centrepoint lies between them cannot be related in terms of proximal and distal, nor can these terms be applied when the lines connecting these points to the functional centrepoint follow different routes. It is therefore not possible to describe the relative positions of the head and the hand in terms of proximal and distal because the line connecting the head to the functional centrepoint follows a different route from that connecting the hand to the functional centrepoint. The terms 'proximal' and 'distal' cannot be used to describe the positional relationship between the pelvis and the thorax because the functional centrepoint is situated between them. The same is true for the hand and the foot.

Indicating place (positional) Fig. 52	*Indicating direction (directional)* Fig. 53
Proximal (near the central point of the body)	Proximally (towards the central point of the body)
Distal (far away from the central point of the body)	Distally (away from the central point of the body)

2.4 Joints as Pivots, Switch Points and Levels of Movement

In functional kinetics, the joint is of interest as the place where movement occurs within the body. In perceiving movements, however, analysing them for therapeutic purposes, and giving movement instructions, what we register is the changes in position of 'levers', 'pointers' and 'gliding bodies', as measured by the changes in distance between

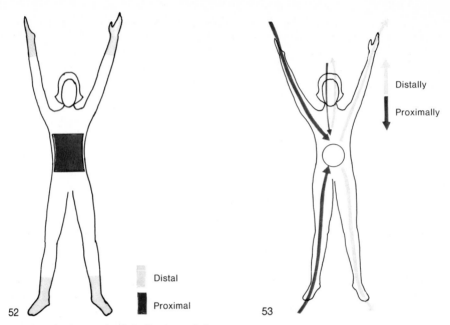

Distally

Proximally

Distal

Proximal

52   53

Fig. 52. Distal – proximal: indications of place

Fig. 53. Distally – proximally: indications of direction

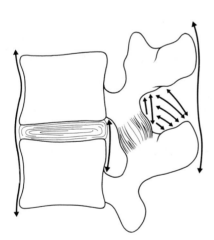

Fig. 54. Motion segment of the vertebral column. (Adapted from Rickenbacher et al. 1985)

points on the body; we register no information about the activities which bring these positional changes about. We speak of the *'fulcrum* or *pivot* of movement', the *'switch point* of movement' and the *'level* of movement'. These designations indicate the difference between the conception of the joint in functional kinetics and that in conventional anatomy.

36

The term 'level of movement' is particularly apt when we consider the lumbar, thoracic, and cervical spine from a functional point of view, because the movements to be defined in these areas always involve the participation of several motion segments. A motion segment of the spinal column consists of two adjoining vertebral bodies, the invertebral disc and the intervertebral joints (Fig. 54).

Note

Movements are displacements of individual parts of the body in relation to each other; they take place at the joints. We shall speak of the fulcra, pivots, switch points and levels of movement and mean those places where movements of the body occur.

To obtain approximate comparative measurements of individual movements the following schematization is necessary: the parts of the body involved in the movement, whether themselves moving or fixed, are reduced to 'levers', 'pointers' and 'gliding bodies' and reference points or axes are mentally projected onto them as required to facilitate observation. The mental image here is of levers turning at a fulcrum about an axis, pointers rotating about a pivot (axis of rotation), and gliding bodies moving across each other in a plane, or their articular surfaces being distracted or compressed. In this way it is possible to observe and measure individual movement excursions or give instructions for them to the patient.

2.4.1 Degrees of Freedom and Components of Movement

A 'degree of freedom' is a way in which a lever, pointer or gliding body can move to and fro at its joint connection. Each degree of freedom has two components of movement – the 'to' and the 'fro'. For instance, the elbow joint has two degrees of freedom: (1) flexion/extension and (2) pronation/supination. A movement at the elbow joint will usually have two movement components: flexion + supination/pronation, or extension + pronation/supination. One way of characterizing joints is by the number of their degrees of freedom, which, doubled, gives the number of their components of movement.

Functional Classification of Joints

Hinge type: 1 degree of freedom, 1 axis of movement, 2 levers meeting at the fulcrum, 2 components of movement.
Rotary type: 1 degree of freedom, 1 axis of rotation, 2 pointers enclosing the pivot and perpendicular to the axis of rotation, 2 components of movement.
Translatory type: 1 degree of freedom, 1 gliding plane, 2 gliding bodies meeting in the gliding plane, 2 components of movement.
Compression and distraction type: 1 degree of freedom, 1 plane of contact, 2 contact bodies meeting in the plane of contact, 2 components of movement.

- *Examples*

Hinge type (hinge joint): proximal and distal interphalangeal joints of the fingers and toes. The components of movement are flexion and extension.

Hinge type with rotary type (rotating hinge joint): the elbow joint with its four components of movement flexion and extension, and pronation and supination. The long axis of the forearm is the axis of rotation and the flexion/extension axes of the hand and elbow joints are the pointers.
The knee joint with its four components of movement flexion and extension, and internal and external rotation. The long axis of the lower leg is the axis of rotation, the flexion/extension axes of the knee and ankle joints are the pointers. Full rotation at the knee joint is only possible in flexion.

Double hinge type (ellipsoid joint): the wrist joints with their four components of movement extension and flexion, and radial and ulnar deviation.

Triple hinge type with rotary type (ball and socket joint): the hip joint with six components of movement of the hinge type, i. e. flexion/extension, frontal abduction/frontal adduction, or simply abduction and adduction; other possible movements are transverse abduction and adduction (i. e. with the long axis of the thigh sagittotransverse). The two movement components of rotation are internal and external rotation.
The shoulder joint with its six components of movement of the hinge type: sagittal flexion and sagittal extension or simply flexion/extension, abduction/adduction and transverse flexion/extension. The two movement components of rotation are internal and external rotation.

Double hinge type with rotary, translatory, and compression and distraction type: the vertebral column has all the degrees of freedom and components of movement listed above. We indicate here only the distribution of the types of movement insofar as this is necessary for reasons of observation and treatment.
Flexion/extension and right and left lateral flexion are possible in all the motion segments of the vertebral column, particularly in the lordotic sections of the lumbar and cervical spine. The levels of rotation are in the cranial lumbar, caudal thoracic, and cervical spine; the scope of rotation is especially extensive at the atlanto-axial joint. Compression and distraction occur in all the motion segments during economical axial loading and unloading of the vertebral column. The levels of movement for extensive and functionally important translatory movements are in the lumbar, lower thoracic and cervical spine.

Note

In normal movement the movement occurring at each individual fulcrum has as many movement components as the joint has degrees of freedom. The range of the different components of movement varies. The anatomical shape of a joint and the topographical organization of its muscles determine the combination of components which will permit economy of performance.

2.4.2 Distance Points

We have already indicated that movement excursions occurring at the joints can be felt by the patient and observed by the physiotherapist as changes in distance between reference points on the body (see p. 53). It follows that we need to establish reference points which are as meaningful to the perceiving patient as to the observing and instructing physiotherapist.

Note

When observing movement at a hinge joint with a fixed fulcrum or a rotary joint with a fixed axis of rotation, we will fail to see the movement excursion of the levers if we keep watching the pivot or axis of rotation. The greatest distance is traversed by those points on levers and pointers which are furthest from the axis of movement. We therefore call these points the distance points of the pivot. For hinge and rotary type joints we have to determine two sets of distance points. They may be proximal and distal or, as the case may be, cranial and caudal to the pivot.

To be able to see a movement at translatory or compression and distraction type joints clearly, we need two distance points on each gliding body or contact body, connected by a line forming an axis.

To define a movement excursion about a fulcrum, the behaviour of the distance point on one moving or fixed part of the body should always be described in relation to the distance point on the other moving or the fixed part of the body.

- *Examples*

Starting position: Sitting upright at a table, the right forearm is placed so that the hand rests on its ulnar border. The long axis of the forearm should be sagittotransverse and the flexion/extension axis of the right wrist should be vertical. The distance point (DP) styloid process of radius moves medially/caudally (inwards/downwards).

Analysis: Pronation of the right forearm effected by the distal pointer.

Starting position: Sitting upright on a stool. The soles of the feet are in contact with the floor and are placed apart in line with the pelvis. The left palm rests on the sternum, the right hand covers the left. The long axis of the sternum performs a parallel gliding movement about 2 cm to the right, the head and arms moving with it, while the pelvis and the legs remain where they are.

Analysis: Translation to the right effected by the cranial gliding body at the level of movement of the lumbar and thoracic spine. The thorax is the cranial gliding body and the long axis of the sternum (formed by the line connecting DPs suprasternal notch and xiphoid process) is the axis of orientation. The pelvis is the caudal gliding body, with the line connecting DPs right and left anterior iliac spines as axis of orientation.

Starting position: Sitting upright at a table; the forearms are placed upon it with the hands resting on the ulnar borders. The long axes of the forearms should be frontotransverse, about 15 cm in front of the rib cage, the right arm in front of the left. The distal distance point of the elbow joint, DP dorsal aspect of the right wrist, moves towards the proximal distance point of the elbow joint, DP right acromion.

Analysis: Flexion at the fulcrum of the right elbow joint effected by the distal lever with pronation of the forearm and external rotation at the shoulder joint, the distal pointer of which – the flexion/extension axis of the elbow joint – has turned in a right lateral direction while the axis of rotation (the long axis of the upper arm) has remained fixed.

In these three examples there is one part of the body that is fixed and one that moves: in the hinge joint there is a fixed proximal distance point and a fixed fulcrum; in the rotary joint, a fixed proximal pointer and a fixed axis of rotation; and in the translatory joint, a fixed caudal and a moving cranial gliding body.

When we want to assess a movement excursion about a joint, we can do so in terms of how the distance points either side of it have moved in relation to each other:

- Does the distance between distance points change?
- In what direction do the distance points move?
- In a hinge joint, how does the fulcrum move?
- In a rotary joint, in what direction does the rotational axis move? Does the axis undergo a parallel displacement? In what plane do the pointers move?
- In a translatory joint, do the axes of the gliding bodies move parallel to each other, or each to itself, or do they deviate, and if so, in what direction?

At the joints of the extremities, movements of the hinge type predominate. The length of the arms can be changed by movement about the flexion/extension axes of the joints of the hand, the elbow and the shoulder; the length of the legs can be changed by movement about the flexion/extension axes of the ankle, knee and hip joints. We can make the arms and legs short or long. It must be remembered that, with a hinge type joint, rectilinear displacement of the distance points (displacement along the shortest route, without deviation) only occurs when the fulcrum moves as well: it moves away from the line connecting the distance points as they approach each other and moves towards it as they move away from each other. In the one instance the fulcrum moves out of the way, in the other it interposes itself between the distance points. If the fulcrum remains fixed, the distance points describe concentric circles with the fulcrum as their centre, while the moving levers represent the radii.

Note

When the fulcrum moves in a hinge type movement, it becomes the distance point of a neighbouring switch point lying in the direction of movement. If the movement is going from distal towards proximal, this is a proximally situated switch point; when it goes from proximal towards distal, it is a distally situated switch point. These switch points must have axes of movement which can be set parallel to the axis of rotation in the moving fulcrum.

Possible Variants of Movements at the Switch Points

Below is a list of the possible variants of movements occurring at hinge, rotary, translatory, and distraction and compression type joints.

40

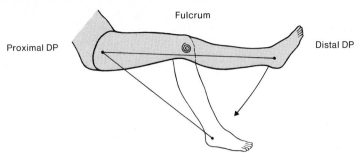

Fig. 55. Flexion at the knee joint (1). The proximal distance point (DP) and the fulcrum are fixed, the distal DP moves

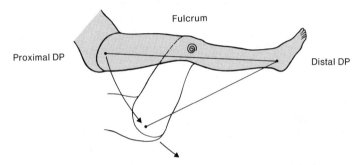

Fig. 56. Flexion at the knee joint (2). The distal DP and the fulcrum are fixed, the proximal DP moves

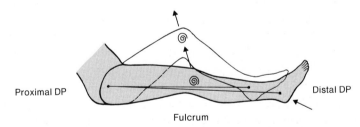

Fig. 57. Flexion at the knee joint (3). The proximal DP is fixed, the distal DP and the fulcrum move

- *Example of the hinge type:* Flexion at the knee joint (Figs. 55–64)

The greater trochanter is the proximal distance point, the lateral malleolus is the distal distance point, and the flexion/extension axis of the knee joint is the fulcrum. We show ten different ways to alter the angle between the long axis of the thigh and the lower leg by flexion at the knee joint: five with the fulcrum fixed and five with it moving. Since flexion involves a shortening of the leg, we see in variants 3 (Fig. 57),

41

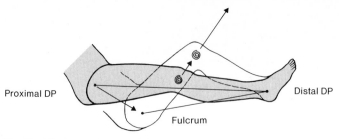

Fig. 58. Flexion at the knee joint (4). The distal DP is fixed, the proximal DP and the fulcrum move

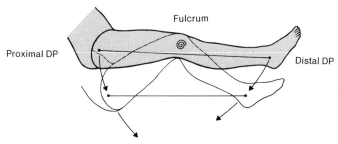

Fig. 59. Flexion at the knee joint (5). The proximal and distal DPs move counter-rotationally, the fulcrum is fixed

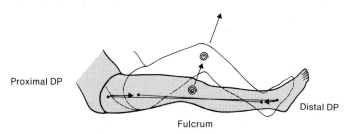

Fig. 60. Flexion at the knee joint (6). The proximal and distal DPs move counter-rotationally, the fulcrum also moves

4 (Fig. 58) and 6 (Fig. 60) how the fulcrum moves away from the line connecting the distance points, which approach each other in a straight line. When the fulcrum is fixed or when it moves in the same direction as the distance points [see variants 1 and 2 (Figs. 55, 56), 5 (Fig. 59) and 7–10 (Figs. 61–64)], the distance points approach each other in a curved line. There are ten analogous variants of extension at the knee joint, in which in every case the distance between the distance points increases and the leg is lengthened.

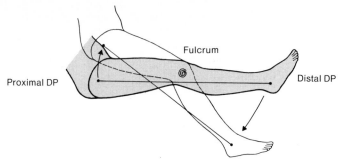

Fig. 61. Flexion at the knee joint (7). The proximal and distal DPs move co-rotationally, the distal DP travelling the greater distance; the fulcrum is fixed

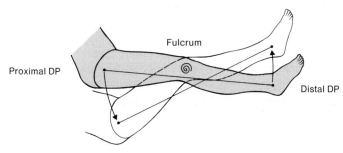

Fig. 62. Flexion at the knee joint (8). The proximal and distal DPs move co-rotationally, the proximal DP travelling the greater distance; the fulcrum is fixed

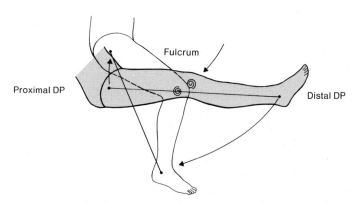

Fig. 63. Flexion at the knee joint (9). The proximal and distal DPs and the fulcrum move co-rotationally; the distal DP travels the greatest distance

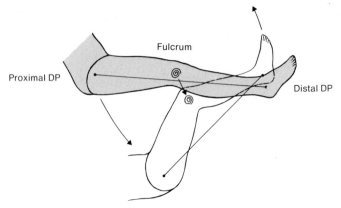

Fig. 64. Flexion at the knee joint (10). The proximal and distal DPs and the fulcrum move co-rotationally, the proximal DP travelling the greatest distance

● *Example of the rotary type:* Rotation at the level of the cervical spine (Figs. 65–69)
The line connecting the earlobes is the cranial pointer and the point of the chin its distance point. The frontotransverse thoracic diameter is the caudal pointer and the suprasternal notch its distance point. The cervical spine is the level of movement and its long axis the axis of rotation. The figures show the five possible variants of a negative (anti-clockwise) rotation of the head in the cervical spine and a positive (clockwise) rotation of the thorax. There are five analogous variants of the opposite rotation in the cervical spine, i. e. positive rotation of the head and negative rotation of the thorax.

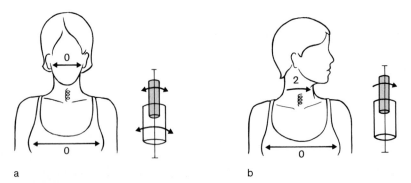

a   b

Fig. 65a, b. Rotation in the cervical spine. **a** Potential directions of movement. The extent of the movement is indicated on a scale of 0–3. **b** (1) The cranial pointer performs a negative rotation. DP point of chin moves dorsally/left, the caudal pointer is fixed

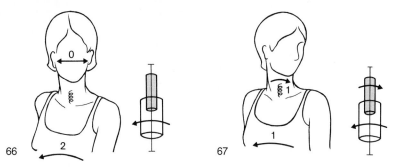

Fig. 66. Rotation in the cervical spine (2). The caudal pointer performs a positive rotation, DP suprasternal notch moving dorsally/right; the cranial pointer is fixed

Fig. 67. Rotation in the cervical spine (3). The cranial pointer performs a negative rotation; DP point of chin moves only half as far as in Fig. 65, going left/laterally/dorsally. The caudal pointer performs a positive rotation; DP suprasternal notch travels only half as far as in Fig. 66, moving right/laterally/dorsally

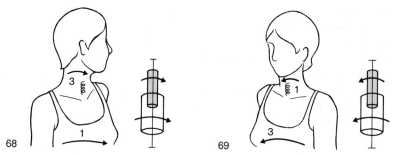

Fig. 68. Rotation in the cervical spine (4). The cranial pointer performs a negative rotation, DP point of chin moving further left/laterally/dorsally than in Fig. 65. The caudal pointer also performs a negative rotation, DP suprasternal notch moving left/laterally/dorsally, but less than the cranial pointer

Fig. 69. Rotation in the cervical spine (5) The caudal pointer performs a positive rotation, DP suprasternal notch moving further right/laterally/dorsally than in Fig. 66. The cranial pointer also performs a positive rotation, DP point of chin moving right/laterally/dorsally, but less than the caudal DP

● *Example of the translatory type:* Lateral movement at the level of the cervical spine (Figs. 70–74)

The line connecting DPs right and left earlobes is the axis of the cranial gliding body. The line connecting DPs suprasternal notch and xiphoid process (the long axis of the sternum) is the axis of the caudal gliding body. The cervical spine is the level of movement. The figures show the five possible variants of translation to the left effected by

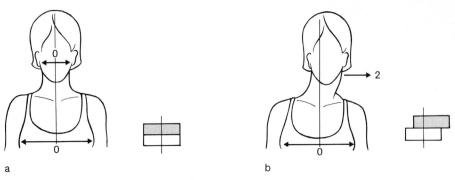

a   b

Fig. 70a, b. Translation in the cervical spine. **a** Potential directions of movement. The extent of the movement is indicated on a scale of 0–3. **b** (1) The cranial gliding body moves left, the line connecting the ear lobes undergoing parallel displacement in its frontotransverse extension; the caudal gliding body, the thorax, remains fixed

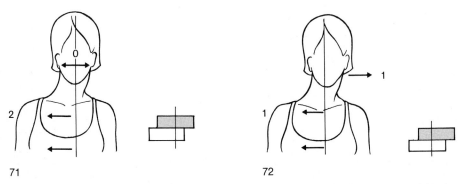

71   72

Fig. 71. Translation in the cervical spine (2). The caudal gliding body, the thorax, moves right, the long axis of the sternum performing a parallel gliding movement to the right. The cranial gliding body, the head, is fixed

Fig. 72. Translation in the cervical spine (3). The cranial gliding body, the head, moves left, but only half as far as in Fig. 70b (1). The caudal gliding body, the thorax, moves right but only half as far as in Fig. 71 (2)

the head and translation to the right effected by the rib-cage; the gliding planes are formed by the motion segments of the cervical spine. There are five analogous variants of translations in the cervical spine in the opposite direction, i.e. to the right effected by the head and to the left effected by the rib-cage. Figures 75 and 76 illustrate all the variants.

46

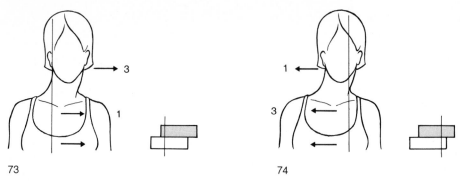

73   74

Fig. 73. Translation in the cervical spine (4). The cranial gliding body, the head, moves left, further than in Figs. 70b (1) and 72 (3). The caudal gliding body, the thorax, also moves left, but not so far as the cranial gliding body

Fig. 74. Translation in the cervcial spine (5). The caudal gliding body, the thorax, moves right, further than in Figs. 71 (2) and 72 (3). The cranial gliding body, the head, also moves right, but not so far as the caudal gliding body

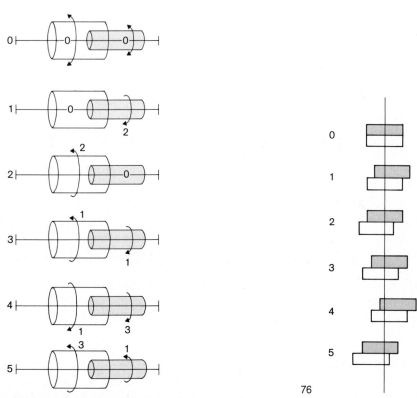

75   76

Fig. 75. The five variants of the rotary type of joint

Fig. 76. The five variants of the translatory type of joint

2.4.3 Influence of Position on Interdependency of Movements at the Joints

Because the long axis of the body is identical with the long axis of the spine and is positioned in the symmetrical plane, and owing to the close positional relationship between the pelvis and the hip joints and between the shoulder girdle and the shoulder joints, the paired switch points of movement of these joints are functionally dependent on each other. The effect of this is to ensure co-ordination of movements of their levers, pointers and gliding bodies.

Note

Movements at the hip joints are automatically transmitted to the vertebral column, particularly the lumbar region.

Movements of one leg at the hip joint are transmitted to the other hip joint and to the vertebral column. The pelvis is set in motion and moves at the other hip joint, while the shape of the vertebral column alters correspondingly within the limits of the available movement tolerance.

Movements of the vertebral column, particularly of the lumbar spine, are transmitted to the hip joints. Movement of one arm at the humeroscapular joint is transmitted to the shoulder girdle on the same side, i.e. to the acromio- and sternoclavicular joints.

Movement of the shoulder girdle at the sternoclavicular joint is transmitted to the humeroscapular joint on the same side.

Since the shoulder girdle, unlike the pelvic girdle, is intrinsically mobile, each humeroscapular joint can alter its position relative to the thorax independently of the other with the aid of movements of the shoulder girdle at the sternoclavicular joint.

The reciprocal dependence of the movements of the arms and the vertebral column is less pronounced than that of movements of the legs and the vertebral column. This is due to the greater movement tolerance at the joints of the shoulder and shoulder girdle and the independence of the movements of one arm from those of the other.

There is thus close functional interdependency between movements of the joints of the vertebral column and those of the proximal joints of the extremities. By contrast, the distal joints of the extremities, particularly those of the upper extremities, are far more independent.

The *frontotransverse, sagittotransverse* and *frontosagittal* lines of intersection of the planes of the three-dimensional system form the axes of movement of the joints of the vertebral column and the head and the proximal joints of the extremities.

2.4.4 Movement About Frontotransverse Axes

Definition: The frontotransverse axes of movement of the vertebral column and of the proximal joints of the extremities are the lines of intersection of the frontal and transverse planes of the body which pass through the centre of the switch points (joints).

Note

Movements about frontotransverse axes are called flexion and extension; the levers move in sagittal planes (Figs. 77, 78). These are hinge type movements.

Exception: Movements at the sternoclavicular and acromioclavicular joints about the approximately transverse long axis of the clavicle are called ventral rotation and dorsal rotation. DP acromion moves ventrally/caudally and dorsally/cranially. These are rotary type movements.

If, in the upright neutral position, the joints of the hips and the vertebral column are flexed, the distance points move forwards ventrally into the field of vision. The distance points of the caudal lever move forwards/upwards (ventrally/cranially) and those of the cranial lever move forwards/downwards (ventrally/caudally), while the fulcrum is displaced backwards (dorsally).
If the joints of the hips and the vertebral column are extended from the neutral position, the distance points move backwards (dorsally) away from the field of vision. Those of the caudal lever move backwards/upwards (dorsally/cranially) and those of the cranial lever move backwards/downwards (dorsally/caudally), while the fulcrum is displaced forwards (ventrally).

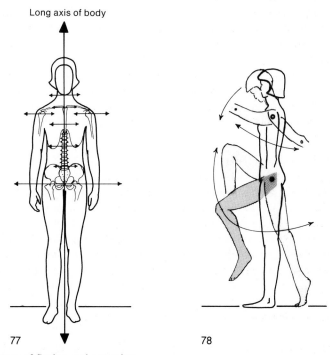

Long axis of body

77   78

Fig. 77. Frontotransverse axes of flexion and extension

Fig. 78. Movements of levers about frontotransverse axes in sagittal planes

49

It follows from this that, in a correct movement instruction, we have no choice as to whether we ask the fulcrum to be moved or the distance points – only one will be right for that movement.

> **Note**
> If, in a gait training session, an extensional movement is called for at the hip joint of the standing limb during a step forward, we can only instruct that the fulcrum be moved forwards. We cannot order a movement forwards with an instruction to move distance points backwards.

2.4.5 Movement About Sagittotransverse Axes

Definition: The sagittotransverse axes of movement of the joints of the vertebral column and the proximal joints of the extremities are the lines of intersection of the sagittal and transverse planes of the body which pass through the centre of the switch points (joints).

> **Note**
> Movements about the sagittotransverse axes of the proximal joints of the extremities are called abduction and adduction; at the joints of the vertebral column they are right and left concave lateral flexion. The levers involved move in frontal planes (Figs. 79, 80). These are hinge type movements.
>
> *Exception:* Movements at the sternoclavicular and acromioclavicular joints about the sagittotransverse axis are called cranialduction and caudalduction. They cause elevation and depression of the shoulder. DP acromion moves cranially/medially and caudally/laterally. These are hinge type movements.

Abduction and lateral flexion at the hip joints and those of the vertebral column from the upright neutral position cause the distance points to move right or left (laterally): the distance points of the caudal levers move right or left and upwards (laterally/cranially), and the distance points of the cranial levers move right or left and downwards (laterally/caudally), while the fulcrum must be displaced medially for abduction and to the left for, for instance, a lateral concave flexion on the right side.

Adduction at the hip joint from the upright neutral position causes the distance points to move towards the middle (medially): the distance point of the caudal lever moves towards the midline/upwards (medially/cranially) and the distance point of the cranial lever moves towards the midline/downwards (medially/caudally), while the fulcrum is displaced laterally.

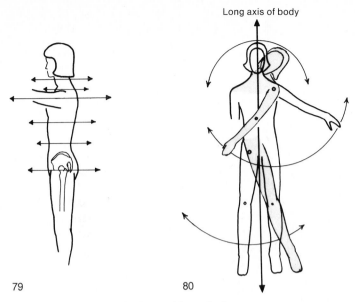

Long axis of body

79   80

Fig. 79. Sagittotransverse axes of abduction, adduction and lateroflexion

Fig. 80. Movements of levers about sagittotransverse axes in frontal planes

2.4.6 Movement About Frontosagittal Axes

Definition: The frontosagittal axes of movement of the joints of the vertebral column and the proximal joints of the extremities are the lines of intersection of the frontal and sagittal planes of the body which pass through the centre of the switch points (joints).

> **Note**
> Movements about the frontosagittal axes of the proximal joints of the extremities are called internal rotation and external rotation; at the movement levels of the spine they are positive and negative rotation. The pointers move in transverse planes (Fig. 81, 82). These are movements of the rotary type. At the proximal joints of the extremities associated hinge type movements take place about frontosagittal axes. At the shoulder joints these are called transverse flexion and transverse extension; at the hip joints they are called transverse abduction and transverse adduction (Figs. 83, 84).
>
> *Exception:* Movements at the sternoclavicular and acromioclavicular joints about the frontosagittal axis are called ventralduction and dorsalduction. They cause forwards and backwards movements of the shoulders. DP acromion moves ventrally/medially and dorsally/laterally. These are hinge type movements.

The line connecting the anterior iliac spines, which lies ventral to the axis of rotation, acts as the caudal pointer in rotation of the vertebral column at the level of the lower thoracic spine, and the frontotransverse diameter of the thorax acts as the cranial poin-

51

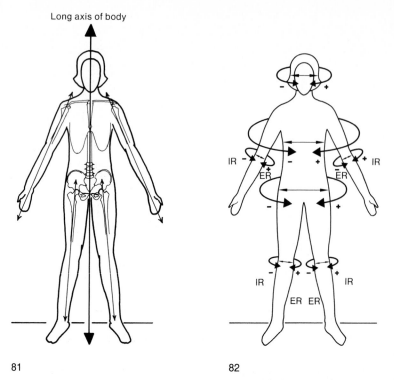

Fig. 81. Frontosagittal axes in positive/negative rotation and internal/external rotation

Fig. 82. Movements of pointers about frontosagittal axes in transverse planes. *ER*, external rotation; *IR*, internal rotation

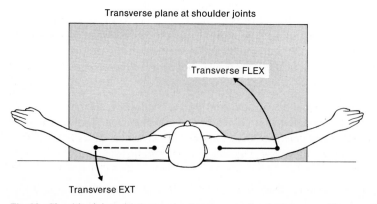

Fig. 83. Shoulder joint with long axis of upper arm frontotransverse. The axes for flexion/extension are frontosagittal

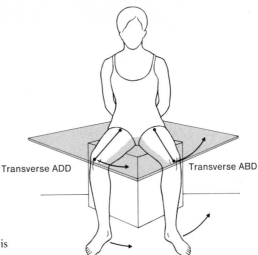

Fig. 84. Hip joint; frontosagittal axes for transverse abduction/adduction. In the neutral position the long axis of the thigh is sagittotransverse

Transverse ADD

Transverse ABD

ter. In rotation at the level of the cervical spine, the frontotransverse diameter of the thorax is the caudal pointer, and the line connecting the earlobes, for instance, could be the cranial pointer. When the statics of posture are normal, these three pointers are parallel to each other and at right angles to the axis of rotation.

To define precisely the quality of the rotation, we speak of positive and negative rotation, that is, clockwise and anti-clockwise respectively, as viewed from above. The level at which rotation takes place and the moving pointer also need to be defined.

- *Examples*

'Positive rotation of the thorax in the thoracic spine' means that the cranial pointer of rotation at the level of the thoracic spine (i. e. the frontotransverse diameter of the thorax) rotates in its transverse plane in a clockwise direction. If it is projected into the plane of the fixed caudal pointer (the line connecting the anterior iliac spines), the angle it makes to its original position can be measured and the extent of rotation expressed in degrees.

'Internal rotation at the right hip joint, effected by the proximal pointer', when the hip joint starts in the neutral position, means that the proximal pointer (the line connecting the anterior iliac spines) rotates in its transverse plane in a clockwise direction. If we project it into the plane of the fixed distal pointer (the flexion/extension axis of the right knee joint), we can measure the internal rotation in degrees.

'External rotation of the right hip joint, effected by the distal pointer' means, in the upright sitting position with the long axis of the thigh horizontal and sagittotransverse, and with 90° flexion at the knee joint, that the distal pointer (the long axis of the lower leg) moves in its frontal plane in such a way that DP right heel moves medially/cranially.

A detailed presentation of axes, distance points, levers and pointers will be found in Chap. 5.

2.5 Capsules and Ligaments of Joints as End-Stops of Movement

If one simply looks at the skeleton of the joints, one would expect fairly large movements from most of them. However, movements are limited by the restraining, 'end-stopping' effects of fibrous capsules, ligaments, muscle tone and muscle activity, which give the actual range or tolerance of movement.

Definition: 'End-stop' is what we call the restriction of the potential mobility of the joints by the passive structures of the motor apparatus.

Note

The location of these end-stops, i. e. where they are in relation to the axes of movement of the various switch points, follows the principle of economy. Muscles which would otherwise have to control possible but unnecessary movements are spared and movements which might endanger the body when it is in motion or when it is held still (e. g. during manual activity) are cut out.

Exception: In muscular contracture, the muscle itself acts as an end-stop, as it has lost its elasticity and become a passive structure.

The degree of restraint varies according to natural constitution and physical condition. When testing the mobility of the joints, the therapist should test movements up to end-stop so as to avoid being misled by braking muscle activity. End-stops are useful safeguards. However, when they are used constantly, they constitute a characteristic of 'bad posture'. There is, indeed, a saving of muscle activity, but the passive structures of the motor apparatus are exposed to unphysiological and uncontrollable strain. The consequences are premature and avoidable wear and tear.

- *Examples*

To assess dorsiflexion at the ankle joint, the knee and toe joints should be flexed (relaxation of the posterior crural muscles and of the long plantar flexors).

The end-stopping of hip joint extension by the iliofemoral ligament at about 15° extension steadies the unstable statics of a human being standing upright and walking. If in standing, however, the ventral end-stops of the hip joints are made to provide the necessary support, the lumbar and lower thoracic spine are subjected to shearing forces, causing deterioration of the vertebral discs (see p. 246).

In genu recurvatum the fully extended knee joint is prevented from giving way by the posterior capsular ligamentous apparatus, without activity of the flexors and extensors at the knee joint. However, the habit of standing with the knee joints overstretched leads to an extensive limping mechanism in which the 'roll-on'[1] phase along the func-

---

[1] Functionally there is no 'push-off' phase in walking, only in running. The term 'roll on' is therefore preferred in reference to walking.

tional long axis of the foot does not take place, encouraging the development of fallen arches and a flat everted foot [see pp. 277–278 and Klein-Vogelbach, *Gangschulung zur funktionellen Bewegungslehre* ('Gait Training in Functional Kinetics')].

Important End-Stops of the Human Motor Apparatus

Foot: The skeleton of the foot, which consists of a great many small bones, has as part of its function to form arches: something which can only be achieved with the aid of end-stops. The dome of the talus – the head of the ankle joint – is enclosed by the double-prong arrangement of the malleoli, which is dependent for its stability on the tibiofibular syndesmosis. Damaged ligaments considerably compromise the function of the foot joints.

Knee: The dorsal end-stops normally prevent the knee from being overextended. Medial and lateral end-stops prevent unwanted abduction or adduction of the lower leg from the thigh. The femoral condyles are prevented from slipping off the tibial plateau by the cruciate ligaments which cross the joint.

Hip: The iliofemoral ligament is one of the most important and most economical end-stops. It prevents the pelvis and the body segments above it from tipping over backwards when the hip joints are extended.

Vertebral column: Each motion segment possesses dorsal, ventral and lateral end-stopping ligaments. They create optimal conditions for the complicated structure of the vertebral column to take axial loading and make it unnecessary, for instance, for flexors to be attached directly to the thoracic spine. The frequency with which postural back trouble occurs clearly shows what happens when undue demands are placed upon the end-stops through bad postural habits (see Klein-Vogelbach, *Therapeutic Exercises in Functional Kinetics,* Chap. 5).

Sternoclavicular and acromioclavicular joints: How important end-stops are is demonstrated by these switch points of movement, which, being the only joint connection between the shoulder girdle and the thorax, have to bear the enormous strain caused by the activities of the hands and arms (see pp. 196–197, 'pincer jaws').

2.6 Muscles as Effectors of Posture and Movement

From the functional point of view, muscles are the effectors of posture and movement. They can act as weight shifters or lifters, they can brake falling weights, or they can prevent weights from falling, i.e. be activated as fall-preventers (Figs. 85–87).

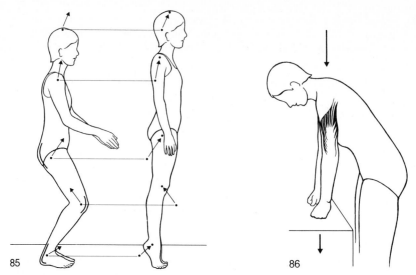

Fig. 85. Triceps surae and quadriceps muscles and extensors of the hip joint as lifters at the ankle, knee and hip joints

Fig. 86. Triceps brachii muscles as preventers of possible flexion at the elbow joint

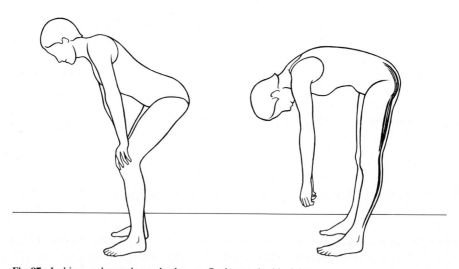

Fig. 87. Ischiocrural muscles as brakes on flexion at the hip joints

2.6.1 Isometric and Isotonic Muscle Activity

The concepts of 'isometric' and 'isotonic' muscle activity are familiar, but these terms are perhaps less than well chosen. It must be remembered that the meaning of 'isometric' is 'staying the same length'; 'isotonic' means 'having the same osmotic pressure' and, in its transferred sense with reference to muscle, 'changing in length'. There are two ways for muscle to actively change in length: 'concentric isotonic', in which the path of the muscle from its origin to its insertion becomes shorter, and 'eccentric isotonic', in which this path becomes longer. (It is possible to alter the path of the muscle without engaging muscle activity, e.g. by manipulation.) It is the path of the muscle that is important, not the distance between its two ends.

Note

We can attach the term *concentric isotonic* to the terms *load shifter* and *lifter*, *eccentric isotonic* to *brake*, and *isometric* to *preventer of possible movement at a joint* (Figs. 88–91).

Concentric isotonic: The muscle shortens actively, shifting and lifting load. In a hinge joint, if the fulcrum is fixed, the distance points perform a circular movement. If the distance points move in a straight line, the fulcrum moves away from this line when they approach each other and towards it when they move apart.
If the angle bridged by the muscle is less than 180°, the distance between the distance points on the levers becomes smaller and the angle they form decreases (Fig. 92). If the angle bridged by the muscle measures more than 180°, the distance between the distance points on the levers becomes greater and the angle they form increases (Fig. 93).
With the axis of movement horizontal, a muscle acting as a lifter exercises positive lift, comparable to the action of the arm of a crane raising a load. Shortening of a muscle against resistance from the therapist is equivalent to this.

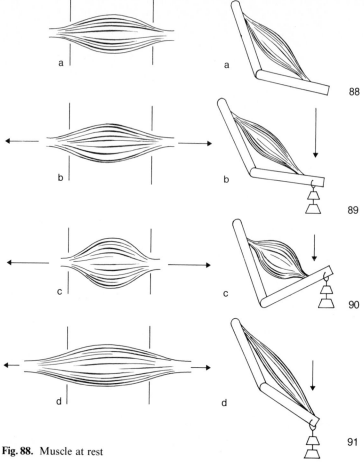

Fig. 88. Muscle at rest

Fig. 89. Active muscle preventing possible movement (isometric): + tension, 0 movement

Fig. 90. Muscle as lifter (concentric isotonic): + movement, − length

Fig. 91. Muscle as brake (eccentric isotonic): + movement, + length

Eccentric isotonic: The muscle actively lengthens, working as a brake on possible movement at the joint. At a hinge joint, if the fulcrum is fixed, the distance points perform a circular movement. If the distance points move in a straight line, the fulcrum moves away from this line when they approach each other and towards it when they move apart.

If the angle bridged by the muscle is less than 180°, the distance between the distance points on the levers becomes greater and the angle they form increases (Fig. 94). If the angle bridged by the muscle measures 180° or more, the distance between the distance points becomes smaller and the angle they form decreases (Fig. 95).

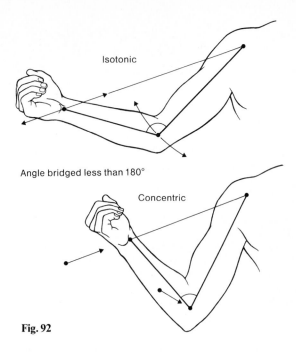

Isotonic

Angle bridged less than 180°

Concentric

Fig. 92

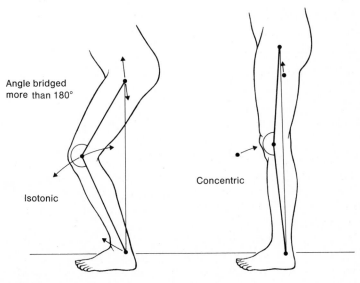

Angle bridged
more than 180°

Isotonic

Concentric

Fig. 93

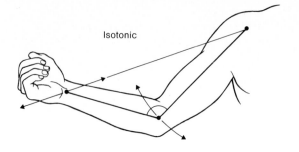

Isotonic

Angle bridged less than 180°

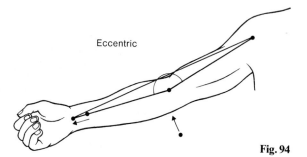

Eccentric

Fig. 94

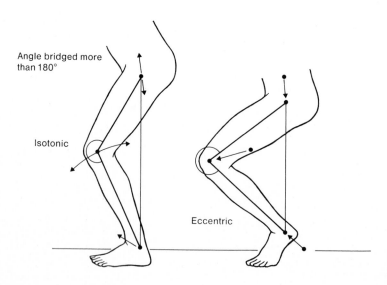

Angle bridged more
than 180°

Isotonic

Eccentric

Fig. 95

60

With the axis of movement horizontal, a muscle acting as a brake on possible move‌ment at the joint exercises negative lift, comparable to the action of the arm of a crane lowering a load. Controlling the yielding of a muscle to a force exercised by the therapist is equivalent to this.

Isometric: The active muscle does not change in length and acts as a preventer of any movement at the joint. If the axis of movement is horizontal and if there is a tolerance of movement downwards, it exercises neither positive nor negative lift. However, it prevents the weight, which would be concentrically raised by the lifter and eccentrically lowered by the brake, from falling. What this muscle achieves is *prevention of falling*. The same static activity is involved in holding a position against resistance exercised by the therapist.

If the axis of movement is vertical, the musculature responsible for initiating movement about this axis does not have to work against gravity, i. e. it can work *lift-free* (see p. 322). Agonists and antagonists both act the same way, i. e. concentric-isotonically, and have to overcome frictional resistance if the lever being moved is resting on a surface. If, when the movement axis of a fulcrum is vertical, the levers are not resting on a base support or supported by the physiotherapist, isometric muscle activity usually has to take place at other joints or levels of movement.

The axis of movement can assume an infinite number of different positions between the horizontal and the vertical. In any movement at a joint, the further the axis of movement inclines from the vertical, the greater the strain imposed on the musculature (see *Therapeutic Exercises,* Sect. 5. 1).

How to remember:
A muscle fulfilling the function described for it in conventional anatomy is working *concentric-isotonically.* The activity of the biceps brachii bends the elbow joint, that of the triceps straightens it (Fig. 96). A muscle performing the opposite function to that described for it in conventional anatomy is working *eccentric-isotonically* (Fig. 97). During a press-up the triceps muscles work eccentric-isotonically as the body is lowered, not straightening the elbow joint but braking its bending. These activities are easiest when carried out against resistance provided by the physiotherapist.

● *Examples*
Training the quadriceps as a *lifter:* Rising from a crouch with the long axis of the body vertical: concentric isotonic training.

Training the quadriceps to act as a *brake:* Bending the knees from standing with the long axis of the body vertical: eccentric isotonic training.

Training the quadriceps to act as a *preventer of possible movement:* Co-contraction of extensors and flexors of the knee joint; the joint is in the neutral position. Here the quadriceps acts as a preventer of flexion, working against the ischiocrural and posterior tibial musculature, while the flexor group acts as a preventer of extension, working against the quadriceps: isometric training. The intensity of this activity can be in-

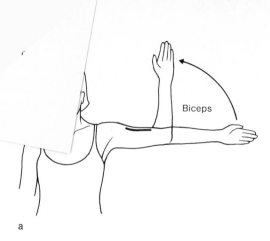

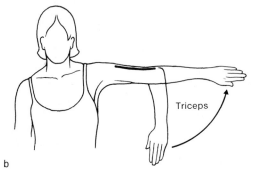

Fig. 96a, b. Concentric isotonic muscle work. **a** Long axis of the body vertical, long axis of the arm frontotransverse, palm facing forwards. DP wrist moves upwards/medially. The flexion at the elbow joint is brought about by concentric isotonic work of the flexors acting as lifters, raising the weight of the forearm. **b** Long axis of the body vertical, long axis of the upper arm frontotransverse, long axis of the forearm frontosagittal/vertical. The hand is under the elbow joint. The extension at the elbow is brought about by concentric isotonic work of the extensors acting as lifters, raising the weight of the lower arm

creased or decreased at will. Co-ordination is important, so that neither extension nor flexion movements take place: the flexors resist the activity of the extensors and the extensors resist the activity of the flexors. This technique is often painless in cases where all movement gives pain; it normalizes the blood circulation and thus improves the movement tolerance.

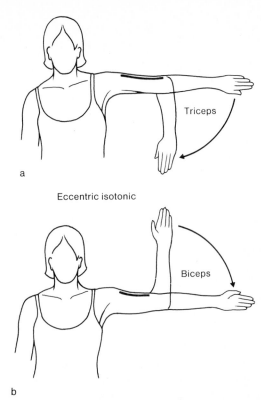

Eccentric isotonic

Fig. 97 a, b. Eccentric isotonic muscle work. **a** Long axis of the body vertical, long axis of the arm frontotransverse. DP wrist moves downwards/medially. The flexion at the elbow joint is brought about by eccentric isotonic work of the extensors acting as brakes, letting down the weight of the forearm. **b** Long axis of the body vertical, long axis of the upper arm frontotransverse, long axis of the forearm frontosagittal. The hand is above the elbow joint. DP wrist moves downwards/left/laterally. The extension at the elbow joint is brought about by eccentric isometric work of the flexors acting as brakes, controlling the lowering of the weight of the forearm

2.6.2 Motive and Compressive Components of Muscle Activity on Joints

Note
Depending on its position in relation to the fulcrum, axis of movement or gliding plane, a muscle can predominantly exert a *motive* effect upon levers, pointers and gliding elements or it can predominantly exert a *compressive* effect upon the articular surfaces of the joint.

Hinge type: The distance between the fulcrum and the line of the muscle's direction of pull (the length of the perpendicular from the fulcrum to the direction of pull) indicates at every position of the levers in relation to each other which of the components

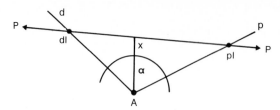

Fig. 98. Motive and compressive components in a hinge joint. *A,* fulcrum; *x,* distance between the fulcrum and the line of the muscle's direction of pull; *P,* direction of pull; *α,* angle between the levers; *d,* distal lever; *p,* proximal lever; *dl,* muscle's point of insertion on the distal lever; *pl,* muscle's point of attachment on the proximal lever

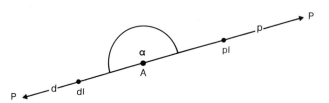

Fig. 99. Motive and compressive components in a hinge joint (abbreviations as in Fig. 98)

is predominant. When the fulcrum and the line of direction of pull are far apart, the motive component predominates; when they are close together, the compressive component predominates. The motive component is particularly strong when the angle between the levers is about 90° and the muscle's points of attachment are as far as possible from the fulcrum. As the angle widens the motive component diminishes and the compressive component increases, and the relative stretching stimulates the muscle to contract. When the angle is about 180° (i. e. when the levers are in extension to each other), the motive component is nil and the compressive component very great, because the muscular contraction presses the articulating surfaces of the joint together (Figs. 98, 99).

Rotary type: When the axis of rotation and the muscle's direction of pull are parallel, the motive component is nil and the compressive component at its greatest. When the axis of rotation is at an angle of about 90° to the muscle's direction of pull, the motive component is at its greatest and the compressive component is nil. Angles from 90° down to about 45° give the motive component greater predominance since the stretching of the muscle is a greater stimulus to contract (Figs. 100, 101).

Translatory type: When the direction of muscular pull is perpendicular to the gliding plane, the compressive component is at its greatest and the motive component is nil. However, as the angle between the direction of muscular pull and the gliding plane becomes smaller, and as the distance between the muscle's points of attachment on the gliding bodies increases, the motive component becomes greater (Figs. 102, 103).
Translatory movements to right and left are important forms of economical motor behaviour, particularly in the regions of the vertebral column between the thorax and the

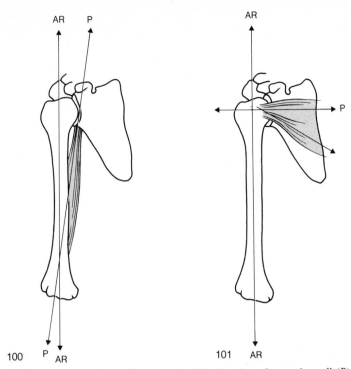

Fig. 100. Motive and compressive components in a rotary joint. The direction of muscular pull *(P)* and the axis of rotation *(AR)* are almost parallel

Fig. 101. Motive and compressive components in a rotary joint. The direction of muscular pull *(P)* and the axis of rotation *(AR)* form an angle between 45 ° and 90 °

pelvis and between the thorax and the head. The musculature of the vertebral column is responsible for these fine movements. It controls stabilization in this area, as illustrated in Fig. 102, and also the movements of the vertebral bodies, which, although small individually, constitute vertebral column movements of some scope when taken all together (Fig. 103). Translatory movement can also be in a ventral or dorsal direction. When the long axis of the body is vertical, all translatory movement in the spine occurs in the horizontal plane, i. e. it is not working against gravity (it is lift-free) and is therefore very economical.

It is known that a muscle exercises the greatest effective force when in the middle position between maximum stretch and maximum contraction. Close to the joints we often find tubercles, trochanters, spines and margins which enhance the motive component.

• *Example*

In the knee joint, the aponeurosis of the quadriceps is separated from the axis of flexion and extension by the patella. This raises the muscle's motive effect in both concentric isotonic extension and eccentric isotonic flexion (Fig. 104 a, b). A compressive component may at any time be added to the isometric activity preventing falling, by simultaneous contraction of the flexors (Fig. 104 c). This always happens when the long

65

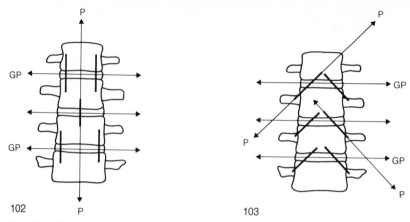

Fig. 102. Motive and compressive components in a translatory joint. The direction of muscular pull *(P)* and the gliding plane *(GP)* are at right angles to each other

Fig. 103. Motive and compressive components in a translatory joint. The direction of muscular pull *(P)* and the gliding plane *(GP)* form an angle of about 45 °

axis of the body is flexed forward from the hip joints, and the ischiocrural musculature controls the tendency of the pelvis to flex at the hip joints by eccentric isotonic or isometric activity. The flexors at the knee joint – the distal fulcrum of the ischiocrural musculature – are then activated simultaneously (Fig. 104 d).

The importance of the compressive component in movements at a joint is undisputed. However, the articular surfaces of the joint need to be compressed by several muscles contracting evenly so that the joint remains stable. This is true for all the joints of the body. The more degrees of freedom available at any switch point of movement, the more complex must be the compressing, stabilizing contractions of the muscles called into play.

● *Example*
The rotator cuff of the shoulder joint. Muscles close to the joint span the capsule and act as compressors, ensuring the coherence of the shoulder joint. The short muscles wrapping the joint facilitate the fine adjustment of rotation and their isometric activity stabilizes the joint during the many skilled movements performed by the hands.

2.6.3 Muscles as Agonists, Antagonists and Synergists

Agonist means contestant. That is an apt name for muscles which are the chief activators initiating movement excursions of levers, pointers or gliding elements at one or – in the case of muscles crossing several joints – more levels of movement. As movers, levers or brakes, agonists work concentric-isotonically or eccentric-isotonically.

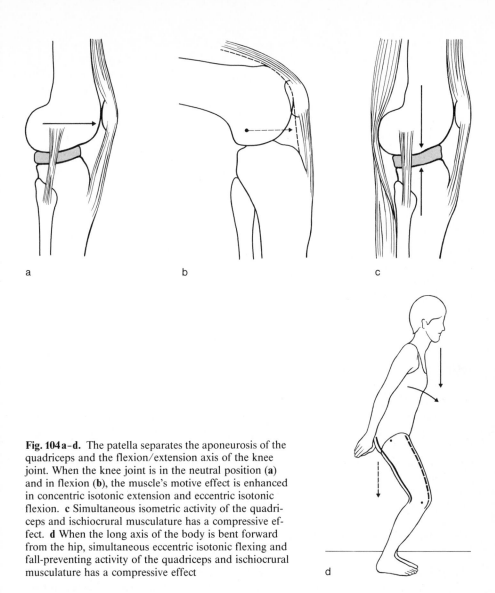

Fig. 104 a-d. The patella separates the aponeurosis of the quadriceps and the flexion/extension axis of the knee joint. When the knee joint is in the neutral position (**a**) and in flexion (**b**), the muscle's motive effect is enhanced in concentric isotonic extension and eccentric isotonic flexion. **c** Simultaneous isometric activity of the quadriceps and ischiocrural musculature has a compressive effect. **d** When the long axis of the body is bent forward from the hip, simultaneous eccentric isotonic flexing and fall-preventing activity of the quadriceps and ischiocrural musculature has a compressive effect

Antagonist means opponent. An antagonist works in opposition to an agonist. If the agonist works concentric-isotonically, the antagonist will work eccentric-isotonically, and vice versa. The antagonist is responsible for harmonizing movement. It regulates the activity of the agonist by inhibiting or enhancing it.

Synergist means partner. Acting on its own, the synergist is less effective than the agonist or antagonist, but it supports both and thus enhances their effectiveness. If the chief activator fails, the synergist can partly fill its role. Synergists may also emphasize only a part-function, i.e. one component of movement of the agonist or antagonist.

Agonists, antagonists and synergists can also act isometrically to prevent possible movement excursions. This is only possible when all the forces acting upon the muscles are in equilibrium. Simultaneous activity of all the muscles crossing a fulcrum is called co-contraction.

- *Example*

The triceps surae muscle is the agonist for plantar flexion at the ankle joint. The flexor digitorum longus, flexor hallucis longus and posterior tibial muscles work medially as synergists; the peroneus longus and brevis muscles work laterally as synergists. The anterior tibial muscle acts as antagonist with the synergists extensor digitorum longus, peroneus tertius and extensor hallucis longus muscles. Thus, for plantar flexion at the ankle joint, we have a group of agonists and a group of antagonists. In the agonist group, multi-joint muscles also activate flexion at the knee and toes; in the antagonist group multi-joint muscles effect extension of the toes.

The upper extremity PNF pattern from flexion/abduction/external rotation to extension/adduction/internal rotation, with the elbow joint at neutral in regard to flexion/extension, shows an example of isometric work of agonists and antagonists at the elbow joint. Balancing these forces demands finely co-ordinated prevention not only of flexion but also – where there is cubitus recurvatus – of hyperextension too (Fig. 105).

From the point of view of functional kinetics, simultaneous activation of agonists, antagonists and synergists is particularly effective in the treatment of muscular disorders. During a joint movement or – as masterfully shown in the PNF techniques – sequence of movements, resistance is selectively applied to neutralize the effect of gravity on the therapeutically important fulcra. A single, clear force of resistance makes the movement more primitive but provides a good opportunity to practise activities which otherwise, because of pain, restricted mobility or muscular weakness, are distorted by avoidance mechanisms to evade the influence of gravity.

2.6.4 Passive and Active Muscle Insufficiency

Definition: A muscle is *passively insufficient* if it cannot be stretched far enough to allow levers, pointers or gliding bodies at the level of movement it bridges to move until end-stopped.

Passive insufficiency may be *physiological,* when a muscle bridges more than one level of movement. Passive insufficiency within the norm is a desirable, economical brake on movement.

Abnormal passive insufficiency occurs when the braking effect sets in prematurely and ̄ movement and statics of posture, often to a considerable degree, ɦortening of the ischiocrural musculature. Passive insufficiency of ses only one switch point is always pathological. It causes changes nomical loading of the passive structures, and changes in the tone ɡ musculature (see Chap. 6).

Fig. 105. PNF pattern of extension/adduction/internal rotation to flexion/abduction/external rotation in the right shoulder joint

Passive insufficiency may be *constitutional.* In this case it can be normalized or at least improved through controlled stretching.

Passive insufficiency may be *conditional,* i. e. caused by poor physical condition. This is usually reversible and can be rectified with relative ease through stretching.

Passive insufficiency may be *pathological.* Such cases often involve muscular contractures, which can only be rectified with great difficulty, if at all.

Definition: A muscle is *actively insufficient* if it is unable to contract strongly enough to actively fix at the limit of their movement tolerance the levers, pointers or gliding bodies at the level of movement it crosses.

Active insufficiency means that the muscle is too long in relation to the parts of the skeleton to which it is attached. Active insufficiency of a muscle bridging only one joint is always *pathological.* If a muscle crosses more than one level of movement, maximum contraction can never be reached over all the joints at once, because of the physiological passive insufficiency of multi-joint muscles. As a rule, the main function of such muscles is at their distal switch points. The multi-joint muscle's ability to contract can be considered normal if it can actively keep the distal fulcrum or fulcra at end-stop even when there is moderate stretching at the proximal fulcrum.

Active insufficiency may be *constitutional* or *conditional,* where there is hypermobility of the joints due to deficient end-stopping (genu valgum and recurvatum, cubitus valgus and recurvatus, etc.), or in cases of chronic physical inactivity.

Traumatic active insufficiency may be seen, for instance, following fractures of the long bones, fractures at the neck of the long bones with displaced major tuberculae, or after patellectomy, partial tearing of muscles, or tearing at myotendinous junctions.

Economy of Multi-joint Muscles

The way in which multi-joint muscles perform their work displays to perfection the economical principles inherent in natural movement. If we think how active and passive insufficiency affect each other (see p. 68), and remember how a muscle exercises the greatest moving or braking force when it is in the middle position between maximum stretch and maximum shortening, because the motive component (see p. 63) and the stretch factor then complement each other, we see how the principle of economy underlies the behaviour of multi-joint muscles.

Note

As distal shortening occurs in a multi-joint muscle, a compensatory lengthening takes place proximally, and the optimum overall length is thus maintained.

Multi-joint muscles have their main function at the distal switch points, where they develop the greatest lifting and braking power.

The most efficient operation of multi-joint muscles is distal concentric isotonic shortening with simultaneous proximal compensatory lengthening. Obviously, there also exists the alternative of distal eccentric isotonic lengthening with simultaneous proximal compensatory shortening.

If, during therapy, one needs the proximal motive component of a multi-joint muscle to function as a 'replacement mechanism', this can be enhanced by stretching the muscle distally.

To train a multi-joint muscle in strength and skill in its main function, i. e. at the distal switch point, we reduce its stretch proximally.

Norms

To evaluate active and passive insufficiency of the musculature, precise knowledge of the norms is required. The norms for stretching and shortening of multi-joint muscles are set out below.

Maximum stretching	**Maximum shortening**
Ischiocrural musculature	
With maximum extension at the knee joint, the hip joint can be flexed to 90° (Fig. 106).	With maximum extension at the hip joint, the knee joint cannot be actively flexed to end-stop.
Rectus femoris muscle	
With maximum extension at the hip joint, the heel can be pressed against the buttocks (Fig. 107).	With maximum extension at the knee joint, the hip joint can be actively flexed to about 90°.
Triceps surae muscles	
With maximum extension at the knee joint, the ankle joint can be dorsiflexed to about 20° (Fig. 108).	With maximum plantar flexion of about 50° at the ankle joint, the knee joint cannot be actively flexed to end-stop.

70

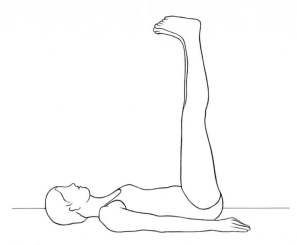

Fig. 106

Fig. 107

Long flexors of the fingers

With maximum extension at the distal, proximal and metacarpophalangeal joints of fingers II–V, the wrist can be extended 70°–80° (Fig. 109).

With maximum flexion at the distal, proximal and metacarpophalangeal joints of fingers II–V, active flexion at the wrist is limited to about 20°–30° (Fig. 110).

Long extensors of the fingers

With maximum flexion at the distal, proximal and metacarpophalangeal joints of fingers II–V, the wrist can be flexed 20°–30° (Fig. 110).

With maximum extension at the distal, proximal and metacarpophalangeal joints of fingers II–V, the wrist can be actively extended to about 70°–80° (Fig. 109).

Pectoral muscles

With the hip joints and lumbar and thoracic spine in the neutral position, the long axis of the arm can be aligned frontosagittally to the cranial projection of the long axis of the body (Fig. 111).

With maximum flexion of the thoracic spine, the upper arm cannot be actively stabilized in end-stopped extension/adduction/internal rotation at the shoulder joint.

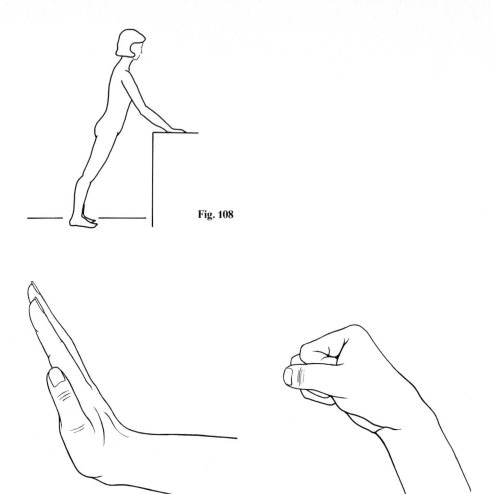

Fig. 108

Fig. 109

Fig. 110

- *Example*

An attacker holding a weapon in his hand will let go of it if the defendant, using both hands in a quick counterblow, forces the attacker's wrist into flexion. The defendant has exploited the active insufficiency of the long flexors of the fingers to loosen the attacker's grasp on the weapon.

Suggestions for Therapeutic Exercises for Muscular Insufficiency

Pathological passive insufficiency can be indirectly treated in accordance with the principle of reciprocal innervation, by maximal loading of the antagonistic musculature. In such cases, even the stretched passively insufficient muscle can serve as resistance. A

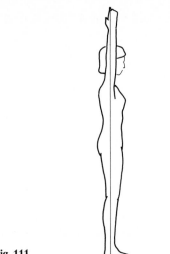

Fig. 111

direct treatment to minimize passive insufficiency is to apply maximum resistance to the affected muscles, with subsequent relaxation.

Pathological active insufficiency can be treated step by step. Initially we utilize the stretching of the muscle at the proximal fulcrum to contract the muscle at the distal switch point to end-stop. Gradually the proximal stretching is reduced. The goal is active fixation of the distal switch point at end-stop with less stretching (i. e. a shorter overall length of the muscle) more proximally than usual. If the muscle has regained the reserves of strength necessary to function well in natural movement, it will no longer be kept out of the action from the start by avoidance mechanisms.

3 Fundamental Observation Criteria

Definition: Observation criteria are features which can be established through systematic observation and palpation of the human body at rest or in motion; they help in distinguishing between the normal and the pathological. These observation criteria are nothing more than the recurrent phenomena typical of a given posture or movement. Differences due to the variations in constitution and condition of the human body may cause these phenomena to vary in degree, but their essential characteristics should not be affected.

Since this book first appeared in German in 1976, work with functional kinetics has led to new knowledge and refinement of the observation criteria. It is highly desirable that this process of refinement should continue, so that the difficult task of registering, interpreting and treating postural and motor deficiency can be further improved.

Natural movement and posture are harmonious and economical, but they are not at all simple. The basis for the economy of the motor system is its capacity for fine adjustment, and this is very complex.

Pathological posture and movement lack harmony, they are uneconomical, and their capacity for fine adjustment is reduced. Posture and movement go out of control, the strain on the passive structures increases, and inappropriate demands are made of the musculature.

Note

For the purposes of therapy we propose to observe and analyse movement as changes occurring in time: firstly as changes in the position of joints within the body, secondly as changes in the position of the body in space.

3.1 Functional Body Segments

As already mentioned, the analysis of movement and posture in functional kinetics are intended to serve as indicators for therapy. The division of the human body into *functional body segments* has proved a helpful form of schematization.

There are five functional body segments: BS thorax, BS pelvis, BS legs, BS head, and BS arms.

BS Thorax: Region of respiration, contains the heart and the mediastinum (Fig. 112)

Skeletal elements: The 12 thoracic vertebrae, the 12 pairs of ribs and the sternum.
Switch points of movement: The 12 motion segments of the thoracic spine and the 12 paired costovertebral joints.
Adjacent body segments: BS head adjoins BS thorax cranially in the area of spinal motion segment T1/C7. BS pelvis adjoins it caudally at spinal motion segment T12/L1. BS arms adjoins it ventrally at the sternoclavicular joints.
Function in motor behaviour: Since three of the four remaining body segments adjoin BS thorax, it is obvious that BS thorax it is destined to be the stabilizing centre of posture and movement. The movement impulses coming in from the periphery - from the legs via the pelvis, from the arms via the shoulder girdle, and from the head - have to

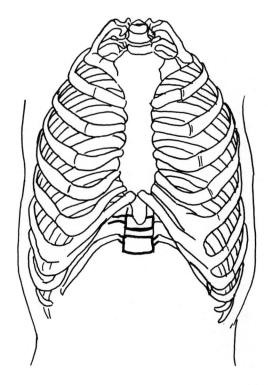

Fig. 112. BS thorax

be stabilized and co-ordinated. The respiratory excursions of the ribs and the positional changes of the long axis of the body in space also demand an adaptable muscle activity, called dynamic stabilization, to maintain the thoracic spine in the neutral position (see p. 86).

BS Pelvis: Region of digestive and reproductive functions (Fig. 113)

Skeletal elements: The five vertebrae of the lumbar spine, the sacrum and the two hip bones.
Switch points of movement: The five motion segments of the lumbar spine, the syndesmoses of the sacroiliac joints and the symphysis pubis.
Adjacent body segments: BS thorax adjoins BS pelvis cranially at spinal motion segment T12/L1. BS legs adjoins it caudally at the hip joints.
Function in motor behaviour: Two body segments adjoin BS pelvis. Being situated between BS thorax and BS legs, the main task of BS pelvis is to achieve a balance between these body segments and their very different functions. The alternating activities of the legs, chiefly in locomotion, need to be 'reined in', co-ordinated and transmitted to the vertebral column by BS pelvis. This enables BS thorax, together with BS head (which is aligned in the long axis of the body), to provide the necessary buttressing (see p. 117) for the activities of the arms. The activity state of *potential mobility* (see p. 87), as observed in the normal erect posture of the human being, enables the pelvis to perform continuous minute balancing acts at the hip and lumbovertebral joints.

BS Legs: Region of locomotion (Fig. 114)

Skeletal elements: In each leg the femur, tibia, fibula, talus and calcaneus; the navicular, cuboid and three cuneiform bones; the five metatarsal and five phalangeal bones.
Switch points of movement: In each leg the knee, talocrural and subtalar joints; the tarsal, tarsometatarsal, metatarsophalangeal and interphalangeal joints.
Adjacent body segments: BS pelvis adjoins BS legs cranially at the hip joints. BS legs has only one body segment adjacent to it; the legs are numbered among the extremities.
Function in motor behaviour: The legs, as the instruments of locomotion, maintain contact with the floor. In locomotion they bring about steady, rhythmical changes of the support area forwards. The legs also form the mobile foundation which supports the vertebral column. Efficient load bearing upon the axes of the legs and feet (see p. 265) is indispensable to sound statics in the vertebral column (see p. 244–245).

BS Head: Region of the organs of sense – eyes, ears, nose, mouth – and of the brain (Fig. 115)

Skeletal elements: The seven vertebrae of the cervical spine and the skull, including the lower jaw.
Switch points of movement: The seven motion segments of the cervical spine together with the atlanto-axial, atlanto-occipital and temporomandibular joints.

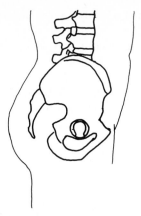

Fig. 113. BS pelvis

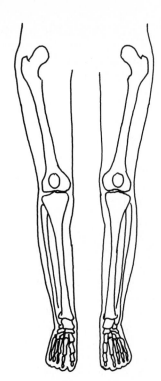

Fig. 114. BS legs

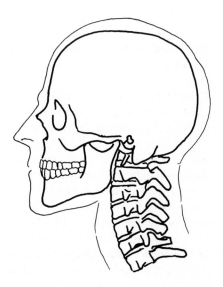

Fig. 115. BS head

77

Adjacent body segments: BS thorax adjoins BS head caudally at spinal motion segment C7/T1.

Function in motor behaviour: BS head adjoins only one other body segment; it is one of the extremities. It balances freely above BS thorax. The delicate control of the positional changes occurring in the vertebral column are regulated from above by the head. To be able to command the whole field of vision and to hear and smell properly, the head needs great potential mobility (see p. 87), which has to be provided by the mobile shaft of the neck.

BS Arms: Region of activity involving manual skills, e. g. writing, drawing, playing music, and the more general activities of grasping, holding, pushing away and leaning (Fig. 116)

Skeletal elements: In each upper limb the clavicle and scapula; humerus, ulna, radius and carpal bones; scaphoid, trapezium and trapezoid; pisiform, lunate, triquetral, capitate and hamate bones; five metacarpals and five phalanges.

Switch points of movement: On each side the acromioclavicular joint, humeroscapular joint and elbow joint; the wrist, the carpal, carpometacarpal and metacarpophalangeal joints, and the proximal and distal interphalangeal joints.

Adjacent body segments: BS thorax adjoins BS arms proximally at the sternoclavicular joints. BS arms has only one body segment adjacent to it; the arms are numbered among the extremities.

Function in motor behaviour: The shoulder girdle attaches the hand to the trunk. It embraces the thorax like a pair of tongs or pincers, forming a fairly loose articulation with BS thorax. The muscular connection between BS arms and the other parts of the body reaches across to BSs head, thorax and pelvis. The arm is a long lever which gives the hand an extensive reach to perform its many varied activities.

How the Five Body Segments Work Together

When we consider the functions of the five body segments and the way in which they work together, we can draw the following conclusions regarding functional motor behaviour:

Three of the five body segments, namely BS legs, BS arms and BS head, are extremities which are attached only proximally to the remainder of the body, their distal ends reaching like tentacles into the environment [see Klein-Vogelbach, *Ballgymnastik zur funktionellen Bewegungslehre* ('Ball Gymnastics for Functional Kinetics'), p. 8]. BS arms and BS legs each consist of symmetrical partners.

BS head is in the plane of symmetry; its neck-shaft is part of the vertebral column. The head has no partner. Raised up on the top of the body, it gathers information from the environment through the senses of sight, hearing and smell. BS head should never leave its position cranial to BS thorax and aligned in the long axis of the body, except when temporarily called upon to do so as part of a particular movement.

If the paired extremities are counted individually, the body has five extremities. The partners *right and left arm* have at their distal ends superlatively skilled tools, the

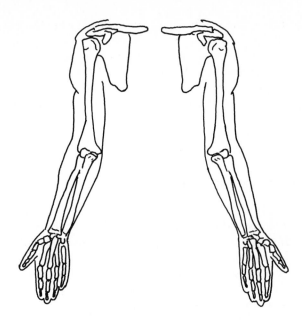

Fig. 116. BS arms

hands. Attached to the body by the long levers of the arms, the hands are in contact with the environment in a multitude of ways and can perform tasks of skill with great precision. All these activities can be executed to a large extent independently.

In locomotion, through their distal ends, the feet, the partners *right and left leg* make alternate contact with the ground and constantly change support area in the direction of locomotion. For this highly co-ordinated and rhythmical process the legs need to have dynamic stability and strength. There is greater interdependence between the legs than between the arms. Because the legs are attached to the stable pelvic girdle, their movements involve BS pelvis.

BS pelvis lies between BS thorax and BS legs. Functionally it can be classed with the lower extremities, since it moderates and transmits the mutually dependent walking movements of the legs to the vertebral column. In walking, it is interesting to note that the mobile lumbar spine, which belongs to BS pelvis, exhibits only slight lateral flexion and hardly any flexion extension, while the activity of the fall-preventing musculature, which prevents movement in the lumbar spine, fluctuates continuously. The rotational component of movement, i. e. the internal rotation of the pelvis at the hip joint and of the femur at the knee joint of the supporting leg, which lengthens the step (see *Gangschulung*), also affects the lower thoracic spine. A remarkable fact is that the dynamic stabilization of the thoracic spine in its neutral position is not adversely affected by this rotation, but rather enhanced by it.

BS thorax functions like a Grand Central Station for the network of movements performed by the extremities. When the thoracic spine is stabilized in the neutral position, it can receive, absorb or transmit all incoming movement impulses. It does this through synergistic or antagonistic muscle activity (see p. 86). The dynamic stabilization of the thoracic spine has to withstand continuous challenges of such diversity that resisting

79

them might well be compared to undergoing physical endurance training. When this stabilizing centre fails to function properly, the disruption of normal function is great (see Sect. 6.7).

We call BS thorax a 'stabile', since it remains unmoving while the other segments, the 'mobiles', move all around it. Sometimes the relationships change, however, as, for instance, in certain activities when BS legs or just one leg temporarily becomes a 'stabile' (Figs. 117–119).

3.2 Activity States

The diversity of possible postures and movements and the body's position in space, subject to gravity, demands various states of activity of the muscles. We will define these states and will then be in a position, when analysing posture and movement, to identify them, relate them to particular body segments, and name them.

To explain how these activity states arise, let us compare the inherently mobile structure of the body to a chain. The links of this chain, joined together by passive structures, can be moved against one another and stabilized in a particular position as and when required. This task is performed by the musculature in its activities of moving, lifting, braking and preventing falling. Such activity states possess typical inherent tendencies to movement depending on whether the body is in contact with a base support, a supportive device, or a suspension device. If we were to observe movement in slow motion, we could analyse it as a sequence of activity states (Fig. 120).

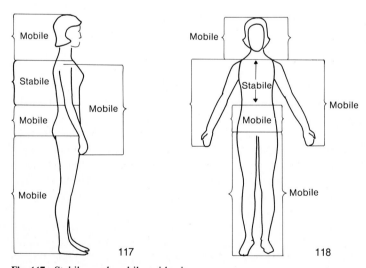

Fig. 117. Stabiles and mobiles, side view

Fig. 118. Stabiles and mobiles, front view

80

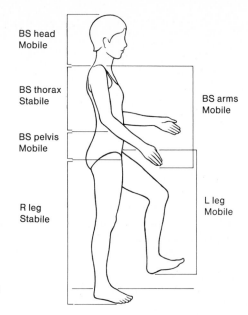

BS head
Mobile

BS thorax
Stabile

BS arms
Mobile

BS pelvis
Mobile

R leg
Stabile

L leg
Mobile

Fig. 119. Stabiles and mobiles in standing on one leg

Fig. 120. Chain

3.2.1 Economical Activity

Definition: When the intensity of the muscular force expended in producing a movement or posture of any kind is neither higher nor lower than what is necessary to achieve the best possible result, the muscle activity is economical. Characteristic of economical activity is that it is pleasing to the eye.

> **Note**
> Muscle activity at very high intensity may still be economical.

If the intensity of muscular activity is too high, it causes stiffness and suppresses the minute changes of position at the joints, particularly in the vertebral column, which constitute the fine equilibrium reactions of the body. Movement becomes clumsy and the strain on critical points is increased. If the intensity of muscular activity is too low,

81

it causes excessive strain and wear and tear on the passive structures of the motor apparatus. Sudden, overcompensating equilibrium reactions occur when the readiness of muscles to react is reduced. Conversely, economical activity increases the speed of reaction in motor behaviour. This is of great importance in physiotherapy: if you wish to train weak muscles in economical activity but at low intensity, the muscles must be engaged only as movers and not as lifters, brakes or fall-preventers (see Sect. 7.3 and *Therapeutic Exercises,* Sect. 5.1).

● *Examples*
By giving appropriate instructions, it is possible to change the intensity of economical muscle activity.

Starting position: If, without any *conditio* having been set (see p. 144), a person raises their arms to the horizontal, the weight of the arms, which has moved forwards, is compensated by a normal equilibrium reaction at the level of the feet by increasing the pressure on the heels, or at the level of the hip joints by a backwards flexing movement of the fulcrum, or at the level of the shoulder girdle by the shoulders moving backwards. Thanks to this equilibrium reaction, the intensity of the economical activity has increased considerably in the hip joints and the shoulder girdle but only slightly in the rest of the body (Fig. 121 a).

If the *conditio* is set that the pressure upon the heels must not be increased and that neither the hip nor the shoulder joints nor the head may move backwards even slightly,

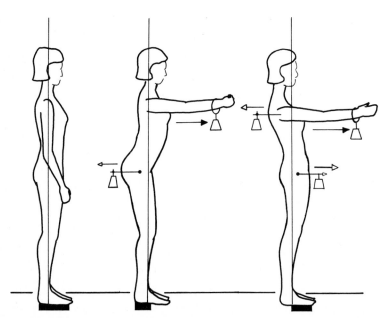

Fig. 121 a. Raising the arms to the horizontal without a *conditio*. Equilibrium reaction backwards at the level of the hip and shoulder joints

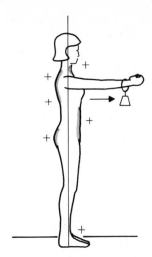

Fig. 121 b. Raising the arms to the horizontal with a *conditio* that there should be no equilibrium reaction backwards; this increases the intensity of the economical activity considerably

the intensity of economical activity increases considerably. In order to fulfil the *conditio*, the equilibrium reactions have to be suppressed, because the weight of the horizontally raised arms may not be evenly distributed over the area of the base support, which would save the body energy. These reactions are suppressed by muscle activity ventrally in the ankle joints and dorsally in the hip joints, the lumbar and thoracic spine, and the region of the shoulder girdle. Then the flexor muscles at the shoulder joints, which have to carry out the movement instruction, are able to lift the weight of the arms to a horizontal position (Fig. 121 b).

> **Note**
> The intensity of economical activity alters according to the weights shifted, lifted, braked or prevented from falling, and also to the speed at which the weights are moved.

3.2.2 Parking Function

The structure of the human body as an inherently mobile system has been compared to a chain (see above). When the chain rests on a horizontal base support and when all its links are in contact with this base, each link has its own centre of gravity and its own base support. We cannot determine a common centre of gravity for the chain as a whole in this instance. Each link is, so to speak, 'parked' on a base support and exerts pressure with its own specific weight upon its base support.

- *Examples*

We decide, let us say, to park the whole body, i. e. all five body segments, on a base support. We will find a completely relaxed, comfortable position (Fig. 122). Since every body segment and every part of every segment has its own base support at the right height, there is no activity against gravity linking the individual parts of the body. The head and neck press upon the pillow, and the thorax, pelvis and upper arms press upon the ground. The forearms press upon the abdomen, the legs upon a supporting pillow. Every part of the body has its own centre of gravity; there is no single centre of gravity and no single support area common to the whole body, as each body segment has its own.

Starting position: Sitting upright on a stool without a back support (Fig. 123). BSs pelvis, thorax and head are aligned vertically in the long axis of the body. BS legs is parked on the ground. The soles of the feet are in contact with the floor and the width of the pelvis apart. The shoulder girdle is parked on the thorax, the arms on the ventral aspect of the thighs. The palms of the hands rest on the thighs with the fingers pointing forwards.

We mentioned that parking function is the activity state with the lowest intensity of economical activity. In our first example (Fig. 122), we can check the completeness of the parking by ensuring that no there is flexional activity (activity of the muscles involved in flexion) at the hip joints and no extensional activity (activity of the muscles involved in extension) at the lordotic section of the vertebral column. To make sure that no adductor activity arises, the legs must be placed such that there is no external rotation at the hip joints. To ensure that the arms are well parked on the abdomen, slight activity is required of the short muscles of the hands to keep them clasped.

In the second example (Fig. 123) we ensure good parking activity for the legs by ascertaining that neither flexional nor extensional nor fall-preventing activity occurs at the

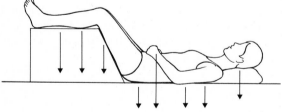

Fig. 122. All the body segments are 'parked' on the base support

hip joints. Fall-preventing potential activity of the transverse abductor and adductor muscles at the hip joints is needed, however, so that the characteristic appearance of an economical positioning of the legs can be maintained.

We started our explanation of the concept of parking function using a chain as our model. As a last point, let us take note that when the body is resting without movement in a completely comfortable, relaxed position, muscular activity between the body segments is disengaged and the body's central point of gravity disappears; each body segment and each individual part of each segment has its own centre of gravity. However, as soon as the spatial relationship between the body segments changes, e. g. when a person starts to get up from the comfortable position, the body segments become linked by muscular activity and the parts thus joined acquire a common centre of gravity. By the time the person has raised him- or herself completely and is standing upright, all the individual centres of gravity of the body segments have united to form a single one for the whole body.

If, in our example in Fig. 123, BS legs is really parked on the floor and the arms on the thighs, BSs pelvis, thorax and head have the same centre of gravity. The weight of the shoulder girdle is a part of this, the thorax serving as its base support. The shoulder girdle is parked on the thorax because its muscular connection to the thorax is inactivated. At the slightest movement, however, this situation changes instantly. For instance, when the long axis of the body is bent forward, the pelvis becomes linked by extensional activity to BS legs, or, when the long axis of the body is bent backward, the legs are attached by flexional activity to the pelvis.

Rising from a sound and relaxed night's rest might be accompanied by the rousing cry 'Centres of gravity of the body, unite!'

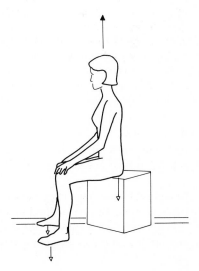

Fig. 123. BS legs parked on the ground, shoulder girdle parked on the thorax, arms parked on the thighs

3.2.3 Dynamic Stabilization

Definition: Dynamic stabilization occurs when one or several joints are fixed in a position by muscular activity.

Note

Dynamic stabilization is an activity state frequently present in economical motor behaviour. Potential mobility and dynamic stabilization are interdependent and can only be understood in relation to each other.

In the economical upright position, the thoracic spine is dynamically stabilized in neutral.

When a level of movement possessing more than one degree of freedom is dynamically stabilized, one of these may be excluded from the muscular fixation and retain its potential mobility.

- *Examples*

When the thoracic spine is stabilized in neutral, the rotation component is potentially mobile. This is because rotary movement in the lower thoracic spine in the erect posture does not have to work against gravity and is needed for the alternating movements of the arms and legs in walking.

When a person polishes a table, for example, there are rapid flexion and extension movements in the elbow joint of the polishing hand, while flexion/extension and radial/ulnar abduction in the wrist and pro-/supination in the forearm are in dynamic stabilization. Some of the movement components of the shoulder are stabilized, depending upon the position of the person polishing. When the polishing movement goes to and fro from left to right, the humerus rotates in the shoulder joint and the flexion/extension and abduction/adduction components are in dynamic stabilization.

The dynamism of the stabilization is especially evident when the joints stabilized:
- Alter their position in space during the performance of a movement sequence
- Are subjected to a speeding up or slowing down of the movement being performed
- Are activated by new movement impulses coming from within or without the body and have to maintain their stabilization against these impulses.

- *Examples*

Arm movements at the shoulder and shoulder girdle joints with the elbow joint dynamically stabilized in given position of flexion.

Making forwards punching movements with a clenched fist when standing upright tends to cause flexion of the thoracic spine. The stabilization of this region must be able to resist such impulses. *The dynamic stabilization of the thoracic spine in the neutral position is exposed to the strongest destabilizing challenges in economical motor behaviour.*

3.2.4 Potential Mobility

Definition: 'Potential mobility' is the name given to the readiness of muscles to respond to stimulation with movement, i. e. changes of position of the joints. As we said above, potential mobility is intimately bound up with dynamic stabilization and the two terms can only be understood together.

Note

When the whole body is in a completely comfortable resting position with all the body segments in parking function, very few equilibrium reactions are necessary, the muscles are unlikely to be called upon to respond to stimulation, and potential mobility is more or less at zero. In this activity state we cannot speak of a single support area for the whole body, but only of individual base supports of body segments at their points of contact with an underlying surface (see p. 89 and *Ballgymnastik,* passim).

When the body segments are linked by muscular activity and the area supporting the body is large, the potential mobility in economical movement is relatively small. The smaller the support area, the more potential mobility is needed in economical movement, because the instability of the body's equilibrium demands a considerable degree of muscular reactivity.

In upright, economical posture the lordotic sections of the vertebral column are in the activity state of potential mobility.

Potential mobility is greatest when the switch points of movement involved do not require muscular activity to maintain position against gravity and prevent falling.

The causal relationship between dynamic stabilization and potential mobility is illustrated by the example of upright, economical posture. The potentially mobile cervical spine needs stable underpinning to enable its long axis to be more or less vertical. The thoracic spine stabilized in the neutral position provides this underpinning. The long axis of the latter is also vertical and extends into the cervical spine. In addition, the weight of the head should be neutrally distributed in relation to the horizontal axes of movement of the cervical spine (see p. 246). Being connected to the bony structure of the pelvic girdle at its most caudal motion segment, L5/S1, the potentially mobile lumbar spine of course needs the potential mobility of the pelvis at the hip joints (see p. 245). In standing on both legs, the potential mobility of the pelvis is mainly restricted to flexional and extensional movements at the hip and lumbar vertebral joints. The potential mobility of the pelvis depends upon proper alignment of the axes of the feet and lower and upper legs (see p. 265), because the legs form the substructure which underpins BS pelvis as well as the entire vertebral column. At the same time, the weight of the superstructure upon the lumbar spine, consisting of BS thorax, BS head, and BS arms, should be neutrally distributed in relation to the flexion/extension axes of the hip and lumbar vertebral joints.

Without dynamic stabilization of the thoracic spine, potential mobility is not available at the lordotic cervical and lumbar spine and they lose their economical motor behaviour.

Potential mobility is not restricted to the vertebral column. It is also employed, for instance, at the knee joints when standing, although the fall-preventing activity of the quadriceps is always engaged as well at the same time.

● *Example*
Distribution of weight in relation to the vertical midfrontal plane when standing upright.

Because the ventral (front) and dorsal (back) weights of the head are neutrally distributed above the cervical spine, the latter is potentially mobile.

In the thoracic spine, because the weight ventral to (in front of) the flexion/extension axes of the thoracic spine is greater than the weight dorsal to (behind) them, extensional stabilization of the thoracic spine is necessary to prevent its collapsing in flexion.

The weights above the lumbar spine are equally distributed dorsally and ventrally (in front and behind), making the lumbar spine and hip joints potentially mobile.

The weights carried above the flexion/extension axes of the knee joints are equally distributed in front and behind, although the weight is slightly greater behind. The knee joints are therefore potentially mobile, with the quadriceps exerting slight fall-preventing (i.e. flexion-preventing) activity (Fig. 124).

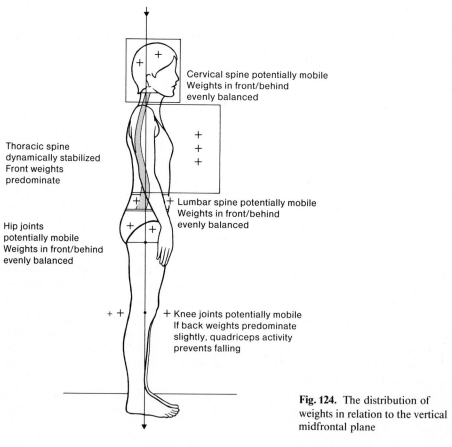

Cervical spine potentially mobile
Weights in front/behind
evenly balanced

Thoracic spine
dynamically stabilized
Front weights
predominate

Lumbar spine potentially mobile
Weights in front/behind
evenly balanced

Hip joints
potentially mobile
Weights in front/behind
evenly balanced

Knee joints potentially mobile
If back weights predominate
slightly, quadriceps activity
prevents falling

Fig. 124. The distribution of weights in relation to the vertical midfrontal plane

3.2.5 Support Area

Definition: The *support area* is the smallest area which includes all the points of contact between activated body segments and their base support.

An important part of our definition is that the body segments are activated. This is because it is by activation that the motor system of the body becomes a single unit with a single centre of gravity, the approximate position of which can be established with reference to the horizontal. If there is only one contact between the body and its base support, the centre of gravity will be found above the point of the greatest pressure. If there are several contacts between the body and its base support and the pressure is evenly distributed, the centre of gravity is over the centre of the body's support area; if the pressure is unevenly distributed, it is located above or close to the point where the pressure is greatest.

- *Examples*

Figure 125 shows the support area in the two-legged stance; the soles of the feet provide the contact with the floor and the functional long axes of the feet point forwards (see p. 265). The weight carried is evenly distributed between the two feet. Figure 126 shows the support area in the one-legged stance; the sole of the foot is in contact with the floor and the functional long axis of the foot points forwards. It is obvious that changing from standing on two legs to standing on one leg reduces the size of the support area by about three-quarters. This explains the considerable alterations which oc-

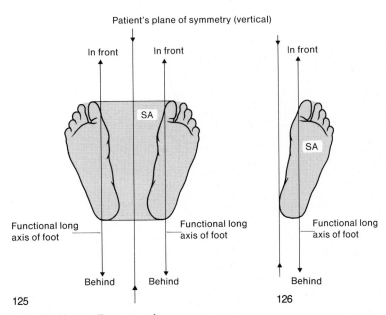

125

126

Fig. 125. Support area *(SA)* in standing on two legs.

Fig. 126. Support area in standing on one leg

cur as the weight is redistributed over the support area, now reduced to a quarter of what it was. It also explains the changes in muscular activity with regard to lifting, moving, braking and preventing weights from falling.

In Fig. 127 it should be noted that the area of contact of the pelvis and the dorsal aspect of the thighs with the bench, projected onto the floor, shows the extent of the support area. So long as BS legs, in contact with the floor through the soles of the feet, remains parked, not linked to BS pelvis by flexional or extensional muscle activity at the hip joints, the main support area is confined to the area of contact between the pelvis and thighs and the bench. BSs pelvis, thorax, head and arms have one common centre of gravity, BS legs has another of its own. However, at the slightest bending forward, the dynamically stabilized BSs pelvis, thorax and head become extensionally linked at the hip joints to BS legs. The support area preprogrammed by the position of the feet when the legs were parked now becomes effective. The legs have changed activity state from 'parked' to 'supporting' and all the body segments now have a single common centre of gravity.

If the patient is seated on a well-inflated gymnastic ball (Fig. 128) instead of a bench, the support area becomes smaller, because it is determined by the area of contact between the surface of the ball and the floor. If the ball is relatively soft, it has a circular area of contact. As the patient's position is very unstable, he will constantly be wanting

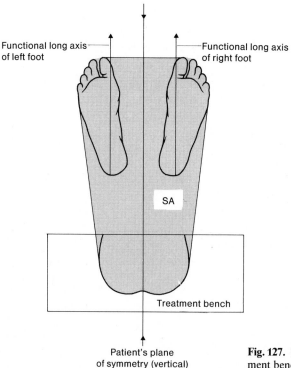

Functional long axis of left foot

Functional long axis of right foot

SA

Treatment bench

Patient's plane of symmetry (vertical)

Fig. 127. Support area in sitting on a treatment bench

to change from having his BS legs in parking function towards having them in incipient supporting function.

Figures 129 and 130 illustrate how the support area changes from that when the patient is on all fours on a box to that when only the left knee and the palm of the right hand are in contact with the box. What has been said about the changes occurring when a two-legged stance is modified to a one-legged stance applies particularly in the present example, because performing this exercise economically demands a state of equilibrium which does not normally occur in our accustomed movement patterns; it is an expressly therapeutic exercise (see p. 144).

3.2.6 Supporting Function

Definition: When one of the extremities is in contact with a base support and exerts more pressure upon this than is due to its own weight, it is in the activity state which we call *supporting function*.

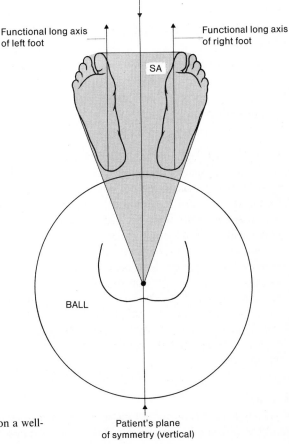

Functional long axis of left foot

Functional long axis of right foot

SA

BALL

Fig. 128. Support area in sitting on a well-inflated gymnastic ball

Patient's plane of symmetry (vertical)

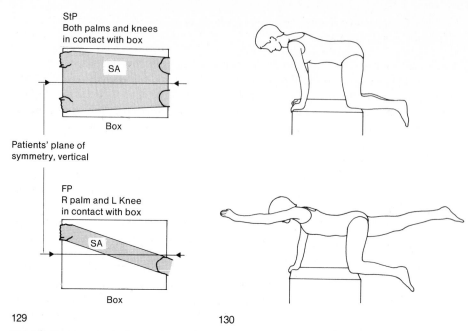

StP
Both palms and knees
in contact with box

SA

Box

Patients' plane of
symmetry, vertical

FP
R palm and L Knee
in contact with box

SA

Box

129

130

Fig. 129 Support area in the starting position *(StP)* of kneeling on all fours with feet unsupported. *FP,* final position

Fig. 130. Support area of the mobilizing exercise into extension from all fours

---

Note

Since the body segments are inherently mobile, they can exert a pressure upon their base support greater than that due to their own weight only when, in addition to the muscles of their joints, the musculature bridging from them to the adjacent body segment is activated in stabilization. We see the phenomenon of supporting function most frequently when the floor is the base support and the pressure applied is in line with the force of gravity; the body's centre of gravity is then above the base support and the joints which are to be stabilized are above each other.

When a body segment is in supporting function, the joints affected are activated against the direction of the pressure exerted by the weight, as if against a continuing movement (see p. 110). The rotational components present in the joints concerned secure the supporting structure by counter-rotational active buttressing (see p. 121) from level to level, each pointer or lever being braced against the rotational tendency of the pointer or lever above it, like a series of lefthand thread and righthand thread screws.

● *Examples*

A person places his or her left hand on a table top for support, thus keeping the right hand free to clean its surface. The left arm is in supporting function and the palm of the left hand is in contact with the base support, in this instance the top of the table (Fig. 131). The support area, therefore, corresponds to the smallest area which includes the areas of contact between the palm of the left hand and the table top, projected on to the floor, and between the soles of the feet and the floor.

The supporting function of the left arm is brought into play because the weight of BS thorax is hanging suspended from the left side of the shoulder girdle (see p. 101) and the pressure produced by this muscular activity is now greater than that exerted by the weight of the left arm alone. The thorax is suspended from the shoulder-girdle by fall-preventing activity of the serratus anterior, rhomboid and trapezeus muscles. The activity of the dorsal extensors prevents palmar flexion of the wrist, extensors prevent flexion at the elbow and abductors prevent adduction at the humeroscapular joint. The screw-type mechanism rotating the forearm as if into pronation, against the external rotation of the humerus in the shoulder joint, acts as a locking device. Without this locking mechanism, the support of the arm is ineffective, as we see in the hemiplegic arm.

Mid-stride position with the weight on the right leg (Fig. 132). The right leg is in supporting function. The area of contact between the right forefoot and the floor is under greatest pressure under the ball of the large toe. Muscles are activated to prevent falling as follows: in the toes, extensional and abductional activity; in the ankle joint, plantar flexional activity; at the knee joint, extensional activity; at the right hip joint, abductional activity, with the pelvis anchored to the right thigh. On the left side the pelvis is anchored to the thorax by lateral flexional activity.

The screw-type locking mechanism comes into play as pronational activity of the forefoot and internal rotation of the tibia in the knee joint counteract the external rotational activity of the femur in the right hip joint, preventing the flexion/extension axis of

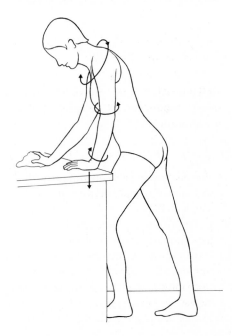

Fig. 131. Increasing the support area when working in the kitchen by having the left arm in supporting function

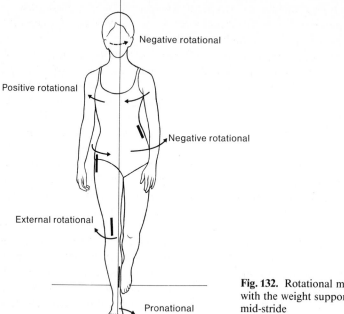

Fig. 132. Rotational movement components with the weight supported by the right leg in mid-stride

the right knee joint from rotating medially. The left–right screwing mechanism continues in the external rotation of the pelvis at the right hip joint, the positive rotational tendency of the thorax, and the negative rotation of the head.

3.2.7 Pressure Activity

Definition: When the pressure exerted by the body upon an area of contact with its base support increases we speak of *pressure activity*. This activity causes compression at the joints involved.

> **Note**
> *Pressure activity* is synonymous with displacement of the centre of gravity towards that area of contact between the body and its base support at which increased pressure is to occur.
> If the pressure is increased on two symmetrical contact areas, the displacement of the centre of gravity is limited to forwards/backwards and up/down.
> If the pressure is increased at a contact area where the body itself provides the base support, alterations of the activity states within the inherently mobile system of the body must take place, allowing the body to redistribute the weight of the body segments.

94

● *Examples*

The patient sits on a treatment bench (Fig. 133). The legs are parked with the soles of the feet in contact with the floor. BS pelvis, BS thorax and BS head are aligned in the long axis of the body, which is vertical. BS arms is parked, the palms of the hands placed symmetrically on the front edge of the bench, next to the hip joints, while the shoulder girdle is parked on BS thorax. In economical activity only the weight of the legs presses upon the areas of contact between the soles of the feet and the ground; only the weight of the arms presses upon the areas of contact between the palms and the bench; only the weight, of BSs pelvis, thorax and head, and of the shoulder girdle presses upon the areas of contact between the right and left ischial tuberosities and their respective areas of contact with the base support.

Instruction: 'The pressure exerted on the bench by your right hand increases considerably, while the sole of your left foot loses contact with the floor and your left hand and left buttock lose contact with the bench' (Fig. 134a). The right leg now takes on a supporting function because the abductors of the hip joint have anchored the pelvis to the right thigh. The weight of the left leg is suspended from the pelvis at the hip joint, the left side of the pelvis from BS thorax, and BS thorax from the shoulder girdle, which is anchored to the right arm at the shoulder joint. The right arm has thus taken on a supporting function. This sequence of movement has brought about a shift of the centre of gravity to the right. BS head has moved with the rest of the body so that the eyes can maintain their horizontal position easily. How the distribution of pressure at the contact areas with the base support (right foot/floor, right hand/bench, right tuberosity/bench) is finally achieved varies between individuals, and can be influenced by the therapist's directing the patient's awareness towards the areas of pressure.

The movement can be changed by giving the following instruction: 'The pressure exerted by your right hand upon the bench increases noticeably while the sole of your right

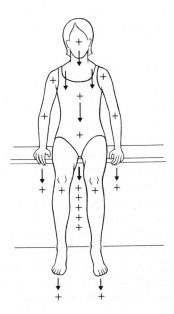

Fig. 133. Sitting on a treatment bench. The long axis of the body is vertical. BS legs is parked with the soles of the feet in contact with the ground. BS arms maintains contact with the front edge of the bench through the heels of the hands

95

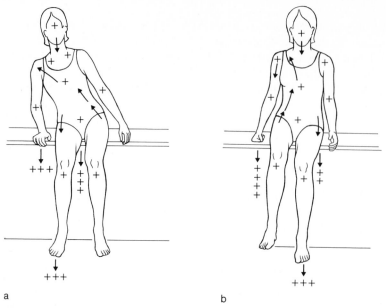

Fig. 134a,b. Pressure activity of the right hand upon the treatment bench

foot loses contact with the floor, and your right buttock and left hand lose contact with the bench.' The left leg now takes on a supporting function because the pelvis has been anchored to the left thigh by the abductors in the left hip joint. The weight of the right leg is suspended from the pelvis at the right hip joint and the right side of the pelvis from the thorax; the thorax is in turn suspended from the right shoulder girdle, which is anchored to the right arm at the shoulder joint. The right arm has thus taken on a supporting function. The left arm is suspended from the left shoulder girdle, which is suspended from the thorax (Fig. 134b).

Instruction from the starting position in Fig. 133: 'The pressure exerted by both hands upon the bench increases considerably but equally on both sides (if the arms are too short, you can make fists and push down on your knuckles). The pressure of your feet on the floor should remain the same and your pelvis should not lose contact with the bench; there should be no increase in muscle activity all around your waist' (Fig. 135). When this instruction has been carried out precisely, the activity states have altered as follows: BS legs has remained parked; BS pelvis has become parked, because the arms, in supporting function, have become a suspension device for BS thorax, which hangs from the shoulder girdle. BS head balances neutrally above this suspension device. BS legs and BS pelvis thus have separate centres of gravity, while the supporting arms, the thorax suspended from the shoulder girdle and the head balancing over the thorax have a joint centre of gravity. With their combined weights they exert pressure upon the base support at the areas of contact the palms or knuckles and the bench.

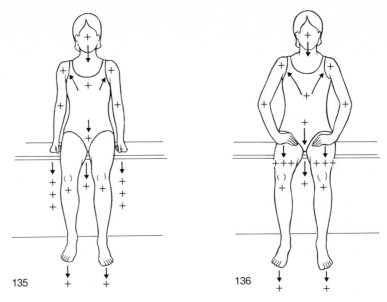

Fig. 135. Pressure activity of both hands upon the treatment bench

Fig. 136. Pressure activity of both palms upon the ventral aspect of the thighs

We alter the starting position shown in Fig. 133. The patient should place the palms of the hands with the finger-tips pointing medially to rest on the ventral aspect for the thighs, close to the hips (Fig. 136). The instruction is: 'The pressure exerted by your hands on your thighs increases considerably; the pressure exerted by the soles of your feet on the floor should just barely increase; the muscle activity in the area of your waist should not increase.' This arrangement alters the activity states within the inherently mobile system of the body but does not alter the pressure exerted at the areas of contact between the environment and the body, nor is the common centre of gravity affected.

3.2.8 Supported Leaning

Definition: When the body is in contact with a base support and leaning against a supportive device, it is in the activity state called *supported leaning*.

> **Note**
> The supportive device is most effective when it is vertical. The supported body segments lean towards the support, for instance, a wall. As the angle of inclination increases, so does the body's tendency to slip at all areas of contact with the environment.

Due to the body's inherent mobility, the topography of the fall-preventing muscles alters when the body leans against a support. The joints in the side of the body facing the support must be braced by the muscles to prevent them from collapsing. When the pressure exerted by walking sticks or crutches does not strike the horizontal base vertically, the sticks cannot provide adequate support – the activity state has changed from supporting (weight bearing) function to supported leaning, and there is a great tendency to slip at the points of contact between the sticks and the floor.

- *Examples*

Figure 137 shows the body in supported leaning against a table. The area of contact is at the level of the hip joints. The long axes of the legs incline forwards at an angle of about 15°. The soles of the shoes are in contact with the floor and the heels of the shoes allow slight flexion at the knee joints. The quadriceps muscles work as preventers of flexion of the knee joint. There is a tendency to slip at both areas of contact between the patient and the environment. In non-slip shoes this form of supported leaning is to be recommended as a relief posture, as it encourages good posture of the spine.

If supported leaning is carried out with the palms of the hands against a wall and the arms, like cross-braces, providing rigid support, the fall-preventing activity is taken over by the flexors of the hip joints and the abdominal musculature (Fig. 138). This posture relieves strain on the dorsal musculature, particularly the lumbar spine, and at the same time provides good training for the abdominal muscles.

Figure 139 shows a patient holding an elbow crutch in her right hand to reduce the strain on the left leg. If the point of contact between the stick and the floor is not under

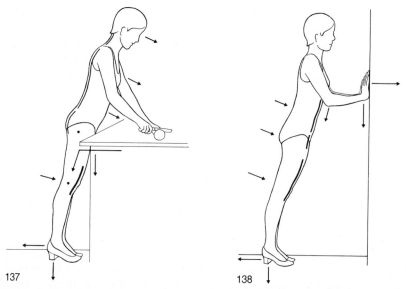

137

138

Fig. 137. The body in supported leaning: contact with the table at the level of the hip joints

Fig. 138. Supported leaning with the hands against a wall

98

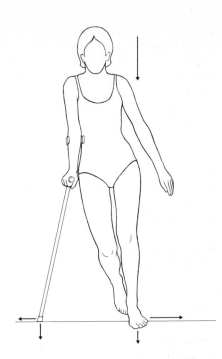

Fig. 139. Poor loading of the long axes of the elbow crutch and the affected leg

the shoulder but too far out to the right, the elbow crutch inclines medially when weight is brought to bear on it and develops a tendency to slip outwards. At the same time, the long axis of the left leg, which is in need of relief, also inclines medially and develops a tendency to slip outwards. The fall-preventing muscles of the medial aspect of the left leg have to be activated to produce adduction at the hip joint and eversion at the subtalar joint. The medial ligamentous apparatus at the knee joint is exposed to uneconomical strain and functional axial loading of the leg (see p. 265) cannot take place.

3.2.9 Free Play

Definition: When an extremity is suspended from the body by its proximal end and is able to move freely at its distal end, it is in an activity state which we call *free play*.

> **Note**
> Body segments in free play are inherently mobile. When they are displaced out of the stable equilibrium they have when hanging, they can only remain freely mobile when their levers are linked by moving, lifting, braking and fall-preventing activities in the muscles bridging the joints. Where two (or more) levers are connected, the more distal is always suspended from the more proximal.

BS arms is made for free play function. The arms are proximally attached to the thorax via the shoulder girdle. In BS legs, the right and left legs are in free play alternately in walking. They are proximally suspended at the hip joint.

BS head is only exceptionally in free play, as when aligned in the vertical long axis of the body in a person standing upright it balances over BS thorax. When the long axis of the body inclines out of the vertical, BS head is in free play in the narrowest sense of our definition, because then it is proximally suspended from the thorax.

When the extremities are in free play and their long axes become horizontal, the state of equilibrium alters considerably, which may affect the pressure on the support area or cause the support area to be changed. The activities opposing the force of gravity take place on the upper aspect of body segments in free play.

Free play function is particularly important for the extremities, especially the arms, as it is the activity state in which the hands carry out tasks requiring skill.

● *Examples*
In Fig. 140 the right leg is in supporting function and the left leg and both arms are in free play.

In Fig. 141 we show the final position of the all-fours stance for mobilizing the vertebral column in flexion and extension (see *Therapeutic Exercises* Sect. 2.1). The right leg, the left arm and the head are in free play. They are proximally attached to the BS pelvis and BS thorax. When their long axes have gained the horizontal, their weight imposes the greatest strain on the fall-preventing musculature.

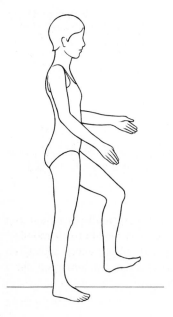

Fig. 140. The right leg is in supporting function, the left leg and both arms in free play

100

Fig. 141. Final position of the all-fours stance in extension

3.2.10 Hanging Activity

Definition: When the whole body, single body segments our parts of body segments are hanging from a suspension device, traction occurs at the joints involved, resulting in what we call *hanging activity.*

> **Note**
> When the whole body or one body segment hangs suspended from either an external device or another part of the body, the parts suspended strive to achieve stable equilibrium.

These hanging weights exert traction upon the switch points of movement involved. The musculature is stretched and thereby stimulated to contracting and lifting activity. When the whole body is hanging by its hands, the levers or pointers at the joints become attached to each other from the most distal to the most proximal and on out to distal again.

● *Examples*
Starting position: Sitting upright on a box (Fig. 142). The hands reach to grasp a bar fixed high above the head. The height of the seat and the bar are such as to allow the weight of BS arms and, hanging from it, BS thorax and BS head, to be suspended from the bar via the hands. The weight of BS pelvis is parked on the box and that of BS legs on the floor.

Starting position: Sitting upright on a box (Fig. 143). With the fingers flexed, the dorsal aspect of the proximal phalanges of fingers II-V forms the area of contact between the arms and the box. By exerting pressure at these points, BS arms goes into supporting function and BS thorax hangs from the shoulder girdle, while the weight of BS head is added to that of the thorax. The weight of BS pelvis is parked on the box and that of BS legs on the floor.

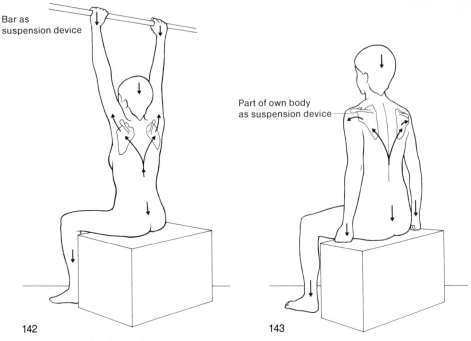

Bar as
suspension device

Part of own body
as suspension device

142

143

Fig. 142. Hanging from a bar

Fig. 143. Thorax hanging from the shoulder girdle

Starting position: Lying supine on the treatment bench (Fig. 144). The right leg per-forms a flexion pattern against resistance exerted by the physiotherapist. The patient's right leg, with its foot dorsiflexed and the toes in extension, hangs from the physiother-apist, who in turn hangs from the patient in such a way as to allow the flexion at the patient's right knee and hip joint to take place. BS pelvis is automatically suspended from BS thorax by activity of the abdominal musculature. The patient's hands find suf-ficient purchase on the edge of the bench to take on a kind of supporting function, causing the thorax to be suspended from the shoulder girdle. This counteracts the body's tendency to slip on the base support.

The skill of the physiotherapist lies in her bringing just the right proportion of her own body weight into play, so that the resistance she provides neutralizes the effect of gravity.

Starting position: Sitting upright on a stool (Fig. 145). A piece of heavy fabric, such as a towel, is draped across the patient's head and the loose ends hanging down serve as a suspension device for the hands. The weight of the arms is suspended from the towel by the hands; the towel, as it were, stabilizes the weight and transmits it as a direct load to the vertebral column. The intensity of the stabilization of the thoracic spine in the neutral position increases slightly. The shoulder girdle is parked on the thorax. The strain on the neck muscles is relieved and the cervical spine and BS pelvis are poten-tially mobile while BS legs remains parked on the floor.

102

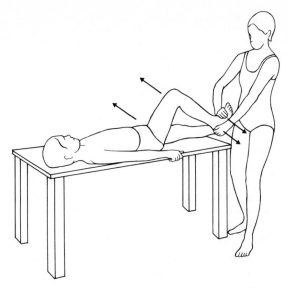

Fig. 144. The patient hangs from the physiotherapist and the physiotherapist from the patient

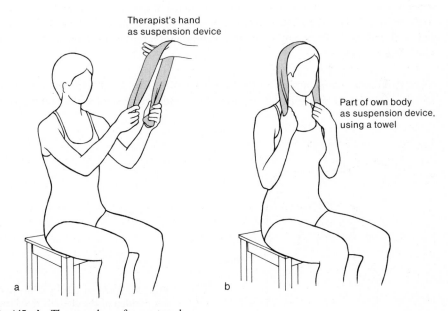

Therapist's hand
as suspension device

Part of own body
as suspension device,
using a towel

a

b

Fig. 145 a, b. The arms hang from a towel

3.2.11 Bridging Activity

Definition: When the body is in a position or carries out a movement in which the support area is defined by more than one point of contact between the body and the base support, the body segments or parts of segments which establish this contact form bridges to the segments adjacent to them. The muscle activity which braces the arches of such a bridge is called *bridging activity.* Since these arches are supported at each end, the lifting force required is small. However, bridging activity can produce a high intensity of economical muscle activity. The muscles bracing the intrinsically mobile arch act as lifters or fall-preventers on the underside of the arch. That the amount of lift required is reduced (see p. 318) results from the fact that there is no free lever, as in free play function, which has to be moved, lifted, braked or prevented from falling (see *Ballgymnastik,* pp. 7–11).

● *Examples*

Starting postition: On all fours on a box with feet unsupported (Fig. 146). The bridge consists of BSs pelvis and thorax: the pelvis hangs extensionally from the thighs and the thorax from the shoulder girdle. The arch is braced by the abdominal muscles, which work as lifters when the lumbar and thoracic spine, the long axis of which should be horizontal, is in too great extension, and as fall-preventers when the thoracic spine is in the neutral position.

Starting position: Sitting on a box (Fig. 147). BS legs is parked on the floor, the soles of the feet in contact with the floor. BS arms, with the heels of the hands on the edge of the box about 30 cm behind the right and left greater trochanters, is in supporting function. BS pelvis and BS thorax form the arch of the bridge, which is characterized by extension of the lumbar and thoracic spine and extensional bracing by the musculature in these regions. BS head balances over the points of contact between the hands and the box, with the cervical spine in a state of potential mobility. The eyes are horizontal and look straight ahead.

Starting position: Lying supine on the floor (Fig. 148). The dorsal aspects of the lower legs are placed on a box about 25 cm high. The hands are clasped and lie on the upper

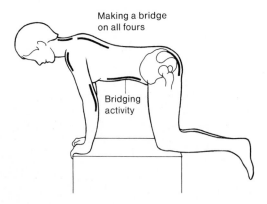

Making a bridge on all fours

Bridging activity

Fig. 146. Bridging activity in the starting position of the all-fours mobilizing exercise

104

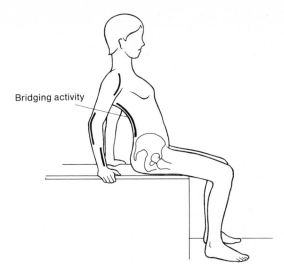

Bridging activity

Fig. 147. Bridging activity of the extensors of the lumbar and thoracic spine in the sitting on a box

abdomen. Through activity of the lower legs pressing upon the box and of the upper arms and back of the head upon the floor, BS pelvis and the caudal part of BS thorax are lifted up off the base support by bridging activity. To achieve this, the extensors at the hip joints and in the lumbar and the lower thoracic spine have acted as lifters. The activity of the head pressing upon the floor has caused the extensors in the cervical spine to form a little arch.

Figure 149 shows a variant of the all-fours mobilization activity in lateral flexion (see *Therapeutic Exercises*) in which there are four bridges. The main one is formed by the left concave lateral flexion of the lumbar and thoracic spine and is supported by the left shoulder and the left greater trochanter. Another bridge is formed by left concave lateral flexion of the cervical spine, supported by the left shoulder and the left side of the head. Adductor activity at the right hip joint causes formation of a bridge for which

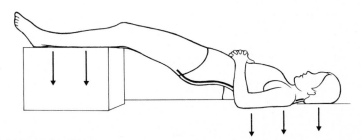

Fig. 148. Bridging activity of the extensors of the hip joints, lumbar spine and lower thoracic spine. Starting position: supine on the floor with the lower legs placed on a box

105

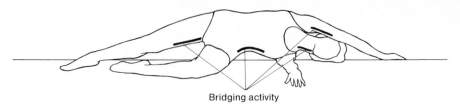

Bridging activity

Fig. 149. Bridging activities in the final position of the all-fours mobilization of the vertebral column in lateral flexion (left-side lying adaptation)

the sole of the right foot and the left trochanter provide the support. Finally, at the right shoulder joint we have an bridge braced by the ventral part of the deltoid muscle and the pectoral muscle and supported by the left shoulder and the palm of the right hand.

Starting position: Standing upright (Fig. 150). BS legs is in supporting function, the feet are in contact with the floor more than a body width apart. The forefeet are in pronation, the ankle joints are in full plantar flexion and the knee joints are in slight flexion caused by minor displacement of the flexion/extension axis forwards. The hip joints flex correspondingly to keep the long axis of the body vertical; the pelvis is potentially mobile at the hip and the lumbar spinal joints, and the head is potentially mobile at the joints of the cervical spine. The thoracic spine is stabilized in the neutral position, while the arms in free play help in maintaining the equilibrium. The bridging activity is in the adductors at both hip joints.

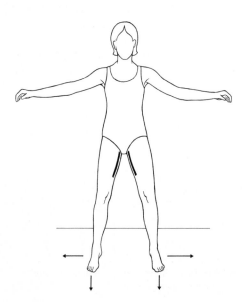

Fig. 150. Bridging activity of the hip joint adductors in a starting position of straddled standing

106

3.2.12 Pushing-Off Activity

Definition: When a surface of contact between the body and a base support or a supportive device is used to achieve purposive thrust, the work done by the muscle is called *pushing-off activity*.

When pushing-off activity occurs, the body's contact with its base support or supportive device is relinquished and the body segment which has performed this push-off automatically goes into free play. The support area becomes smaller, the body's state of balance alters, and the weight of the body segment now in free play has a braking effect in relation to the direction of push-off.

If we intend to use a body segment in free play to push off, it must first come into contact with a base or supportive device.

● *Example*

In going up stairs economically, the lifting work of the right leg is reduced by the left leg's pushing off from the floor and the right arm's hanging on the handrail. The extensors and plantar flexors of the left leg participate in lifting the body's weight up onto the next step; the participating muscles in the left arm are the extensors of the shoulder joint and the flexors of the elbow joint. The joints of the left leg thus undergo compression, those of the left arm traction. As soon as the left foot loses contact with its base support, it is in free play (Fig. 151).

In going up stairs economically, the heel does not touch the base support when the foot is placed on a new step. In this way, firstly, the musculature responsible for plantar flexion at the ankle joint is forced to participate in the lifting, and, secondly, the

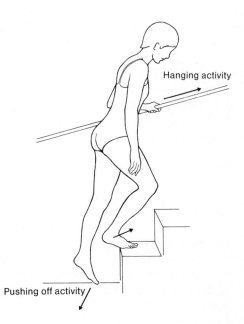

Hanging activity

Pushing off activity

Fig. 151. The left leg pushing off and the left arm hanging from the rail in going up stairs

avoidance mechanism of backward displacement of the flexion/extension axis of the knee joint is prevented, because, with no contact between the heel and the step, this would result in the whole foot's slipping off backwards.

3.2.13 Practical Use of Activity States in Functional Analysis of Starting Positions and Movement Sequences

Because our analysis is therapy-oriented, *economy of activity* is a basic demand, so only the intensity need be specified.

As is well known, too great intensity of activity causes stiffness, while too low intensity overloads the passive structures of the motor apparatus.

When, for instance, the body is leaning against a support, but the muscles on the aspect of the body facing the support are not in the activity state of supported leaning – i. e. they partly or wholly fail to brace the joints they bridge – there is no fall-preventing activity in that body segment, which thus hangs uneconomically from other body segments.

The basic building blocks of economical activity are first, *dynamic stabilization, potential mobility* and *parking function* at the correct joints in the inherently mobile system of the human body and, secondly, proper arrangement of the weights of the body segments *over* the points or areas of contact between the body and a base support, *next to* or *touching* a supportive device, or *under* a suspension device. In other words, economical activity shows itself impressively in the way in which the body is in contact with its environment.

- When the body is in contact with the environment at several points, the distribution of weights is what decides whether support areas are formed, and where (see pp. 89). The critical factor is which weights within the intrinsically mobile system of the body have to be held in a particular arrangement by fall-preventing muscle activity.
- If one of the body's points of contact with the environment is against a supporting device, the body also needs a point of contact which will serve it as a base support. The fall-preventing supported leaning activity on the side of the body facing the support brings about the formation of a support area, as we have described earlier.
- A suspension device external to the body and from which the whole body is suspended makes a support area superfluous.

Free play of the extremities occurs only within the inherently mobile system of the body and requires a base support and/or a supportive or suspension device in the environment. *Supporting* always takes place over a base support, which may be partly formed by the body itself. An important point is that the base support external to the body becomes a support area during supporting activity.

Hanging within the inherently mobile system of the body is triggered when a body segment in free play hangs down perpendicularly or when, in bridging within the body, the arch hangs from its supporting pillars, not braced by muscle activity. An example of this can be seen in a person on all fours with insufficient intensity of muscle activity.

Supported leaning of the inherently mobile system of the body requires not only a supportive device, which may be part of the body, but also a base support. The body has a tendency to slide on the base support, particularly when the supported leaning is carried out uneconomically, since in supported leaning, unlike in supporting function, there is no counter-active screw-type locking mechanism because the long axis of the supported body segment is no longer vertical and the segment therefore cannot be axially loaded. The intensity of economical activity decreases more rapidly in supported leaning than in supporting function because there is little tendency to fall. The dangers of the tendency to slip are not always recognized.

Pressure activity is the way to supporting function.

Pushing off is the way to free play.

Reducing pressure is pushing off in slow motion.

A rapid increase of pressure in one part of the body stimulates a pushing off in a different part of the body or an accelerated horizontal, rectilinear movement of a body segment in free play.

We have described a number of different activity states and are now able to identify them in any given posture or movement of the body. These states can also be called forth in the patient by appealing to his perception.

3.3 Analysis of Movement by Differentiating Between Equilibrium Reactions

Every movement sequence can be understood in terms of a series of equilibrium reactions. In our analyses of movement we must bear in mind that therapy needs to aim at movements produced economically.

In analysing a movement sequence we determine whether it involves *changing location* or maintaining a *constant location*.

Movement sequences which maintain a *constant location* are characterized by the fact that their support area does not change. There are two variants: in one there is no change at the support area at all, while in the other the distribution of the pressure exerted on the surface alters.

Movement sequences which *change location* are characterized by the displacement of the total motor system of the inherently mobile human body from one initial support area to another, new one. If, for instance, the movement is forwards, all (reference) points on the body move forwards. However, parts of the intrinsically mobile system of the body may simultaneously move backwards or upwards or downwards or to the left or right in relation to orientation within the body and outwards from the body.

Note

In observing movement sequences, we are watching the displacement of the weights of body segments or parts of them. If a weight is displaced vertically, the muscles work as lifters and brakes over an unchanging support area. However, as soon as there is a horizontal component to a primary load-shifting movement, it invokes various observable equilibrium reactions, as follows:
1. A counterweight is brought into play.
2. A counteractivity occurs.
3. The support area is changed in the direction of the main weight.
Counterweighting and counteractivating often occur together. Counterweighting with simultaneous change of support area in the direction of the primary weight displacement is usually uneconomical and one of the most common forms of avoidance and limping mechanisms.

3.3.1 Continuing Movement

Definition: When any given point of the body is guided by a movement impulse in a particular direction, and movement excursions take place at the adjacent joints which help to bring about this intended movement, a *continuing movement* has taken place.

Continuing movements are early patterns of motor development. In the motor behaviour of an adult they are always subelements of a movement sequence.
The direction of the movement determines which components at the joints participate in a continuing movement. All the axes of movement which the predetermined direction of movement meets at a favourable angle make their movement tolerance available to the movement impulse. They are, so to speak, in the 'line of fire' of the continuing movement.

Critical Distance Point

The point on the body which most clearly and unambiguously maintains the direction of a continuing movement is its *critical distance point*. It is imperative for the physiotherapist to determine the critical distance point as it is needed for analysis and also, even more importantly, for instructing the patient. Indicating in which direction and how far the critical distance point should move makes it much easier for the patient to carry out a movement instruction.
The critical distance point of a continuing movement is at the same time the distal or proximal distance point of the nearest fulcrum participating in the continuing movement.

● *Examples*
If, when sitting upright, a person bends his head forwards/downwards in an attempt to bring the vertex down to the knees, the vertex is the critical distance point for the entire

movement sequence and at the same time the distal distance point of the upper joints of BS head (the atlanto-occipital and atlanto-axial joints).

When the critical distance point of a continuing movement is situated in the immediate proximity of the fulcrum of a hinge joint which is participating in the continuing movement, a movement takes place at that joint through displacement of the fulcrum. If, for instance, in sitting upright, the attempt is made to move the knee as close as possible to the chin, the patella is the critical distance point for the entire movement sequence. Because of the patella's proximity to the flexion/extension axis of the knee joint, flexion occurs at that joint due to the displacement of the fulcrum and distal DP lateral malleolus on the lower leg.

Critical Fulcrum

The critical fulcrum of a continuing movement is the last fulcrum taking part in that movement. No additional distance point pertaining to a more distant fulcrum which might participate in the movement may be set in motion.

Distance Points of Fulcra Participating in Continuing Movement

Each fulcrum participating in a continuing movement possesses a distal and a proximal or, where appropriate, a caudal and a cranial distance point. Changes in the distance between these distance points provide quantitative and qualitative information about the movement taking place, i. e. information about the *extent of movement excursions* and the *involvement of the movement components* available at the levels of movement concerned.

The extent of the movement excursion is determined by the direction taken by the critical distance point of the continuing movement. Which of the available movement components participate depends upon the position of the axes of movement involved relative to the 'line of fire' of the continuing movement; those which participate are those which are approximately at right angles to the direction of movement and parallel to each other or come to be so during the course of the movement.

Spatial Path of the Critical Distance Point

Basically, the path through space of the critical distance point of a continuing movement can be straight or curved. A straight path may go into a curve and vice versa. However, if a curving path is more like a wavy line, we are no longer dealing with a typical continuing movement, because each wave requires a new impulse. The curving paths of continuing movements go in a circle or spiral inwards or outwards. The length of the levers between the participating fulcra is the decisive factor which determines the spatial path of the critical distance point. The longest levers are in BS legs and BS arms; they are the lower and upper legs and the forearms and upper arms.

Co-rotational and Counter-rotational Continuing Movements

Because of the difference in length between the levers, the distance points of fulcra participating in a continuing movement behave differently depending whether the critical distance point follows a straight or a curving path. The path described by every distance point participating in a continuing movement is determined by two components: by the circular movement about its fulcrum and by the direction of movement of the critical distance point.

If we compare the directions of the circular movements of all the distance points involved in a continuing movement, we see the following possible combinations:

1. All the distance points are moving clockwise along the circumference of their circle (positive rotation), or all are moving anti-clockwise (negative rotation). Their movements are *co-rotational*.
2. The rotation of one distance point about its fulcrum is positive, that of the next is negative, and so on. Their movements are *counter-rotational*.
3. The distance points of a continuing movement move *co-rotationally* (in relation to the fulcrum before) at some fulcra and *counter-rotationally* at others. This we call a *mixed continuing movement*.

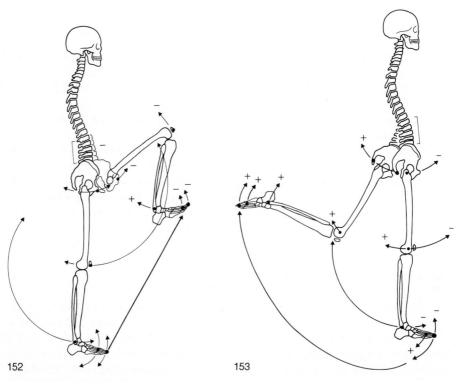

152   153

Fig. 152. Mixed continuing movement from caudal to cranial; direction of movement cranial/ventral; critical DP tips of the toes

Fig. 153. Co-rotational continuing movement from caudal to cranial; direction of movement cranial/dorsal; critical DP tips of the toes

112

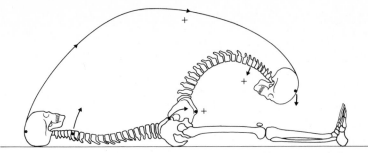

Fig. 154. Co-rotational continuing movement from cranial to caudal; direction of movement caudal/ventral, caudal/dorsal; critical DP vertex

---

Note

In BS arms and BS legs, counter-rotational movements are typical and economical. The arrangement of the multi-joint muscles (see p. 70) also shows this.

In the vertebral column, an increase in the physiological curves, as occurs in hyperkypholordosis, constitutes a counter-rotational continuing movement tending towards compression. A decrease of the physiological curves, as, for instance, in total flat back, constitutes a counter-rotational continuing movement tending towards pull or traction.

● *Examples*

Mixed continuing movement from caudal to cranial (Fig. 152).

Critical DP: Tips of the toes

Direction of movement: Cranially/ventrally

Spatial path of critical DP: Straight line

Critical fulcrum: Lumbar spine

Co-rotational continuing movement: DP tips of toes moves with extension in the toe joints; DP proximal joint of big toe moves with dorsiflexion in the ankle joint.

Counter-rotational continuing movement: DP lateral malleolus moves with flexion in the knee joint and DP patella with flexion in the hip joint.

Co-rotational continuing movement: DP symphysis pubis moves with flexion of the lumbar spine (critical fulcrum).

Co-rotational continuing movement from caudal to cranial (Fig. 153)

Critical DP: Tips of the toes

Direction of movement: Cranially/dorsally

Spatial path of the critical DP: Curving, spiralling first outwards then inwards

Critical fulcrum: Lumbar spine

Co-rotational continuing movement: DPs tips of the toes move with flexion in the toe joints; DPs metatarsophalangeal joints move with plantar flexion in the ankle joint; DP lateral malleolus moves with flexion in the knee joint; DP patella moves with extension in the hip joint; DP coccyx moves with extension in the lumbar spine (critical fulcrum).

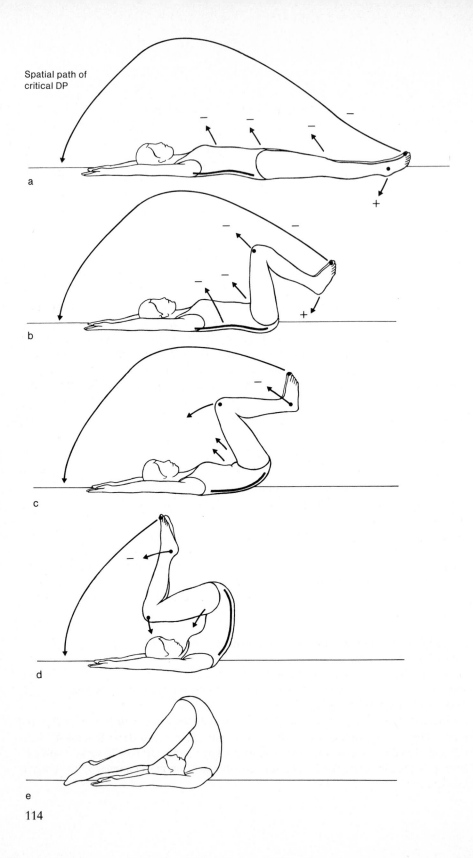

Spatial path of critical DP

a

b

c

d

e

114

Co-rotational continuing movement from cranial to caudal (Fig. 154)
Critical DP: Vertex
Direction of movement: Caudally/ventrally, caudally/dorsally
Spatial path of the critical DP: Curving, spiralling inwards
Critical fulcrum: Hip joints
Co-rotational continuing movement: DP vertex moves with flexion in the atlanto-axial and atlanto-occipital joints and the cervical spine: DP C7 moves with flexion of the thoracic and lumbar spine; DPs anterior iliac spines move with flexion in the hip joints (critical fulcrum).

Mixed continuing movement from caudal to cranial, or from distal to proximal to distal (Fig. 155)
Critical DP: Tips of the toes
Direction of movement: Cranially/upward, cranially/downward
Spatial path of the critical DP: First in a straight line, then curving, spiralling outwards
Critical fulcrum: Thoracic spine
Co-rotational continuing movement: DP tips of the toes moves with extension in the toe joints; DPs metatarsophalangeal joints move with dorsiflexion in the ankle joints.
Counter-rotational continuing movement: DPs lateral malleoli move first with flexion, then, as soon as the heels are over the head, with extension in the knee joints. DPs patellae move with flexion in the hip joints.
Co-rotational continuing movement: DP symphysis pubis moves with flexion of the lumbar spine; DP T12 moves with flexion of the thoracic spine (critical fulcrum)

Co-rotational continuing movement from distal to proximal (Fig. 156)
Critical DP: Tips of the fingers
Direction of movement: Proximally. (Since the hand can assume so many different positions in relation to the body, it is impossible to describe its spatial path in terms of orientation within the body.)
Spatial path of the critical DP: Curving, spiralling inwards
Critical fulcrum: Wrist
Co-rotational continuing movement: DPs finger tips move with flexion at the interphalangeal and metacarpophalangeal joints; DPs proximal phalanges move with flexion of the wrist (critical fulcrum)

Mixed continuing movement from distal to proximal (Fig. 157)
Critical DPs: Tips of the fingers
Direction of movement: Proximally
Spatial path of the critical DPs: In a straight line
Critical fulcrum: Wrist
Co-rotational continuing movement: DPs fingertips move with flexion at the distal interphalangeal, proximal interphalangeal and metacarpophalangeal joints
Counter-rotational continuing movement: DPs metacarpophalangeal joints move with dorsal extension of the wrist (critical fulcrum)

◀───

Fig. 155a–e. Mixed continuing movement from caudal to cranial or from distal to proximal to distal; critical DP tips of the toes

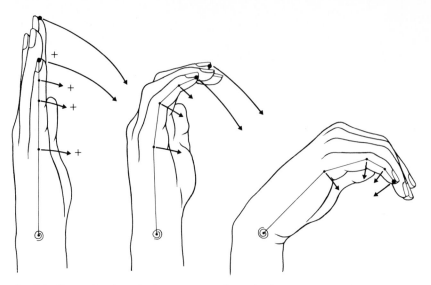

Fig. 156. Co-rotational continuing movement from distal to proximal; critical DP fingertips

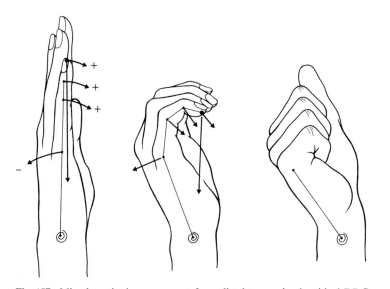

Fig. 157. Mixed continuing movement from distal to proximal; critical DP fingertips

In this discussion of continuing movement we have so far touched upon neither the *speed* of movement nor the *intensity* of muscle activity, because we were leaving the *position in space* of a continuing movement out of account for the time being. However, we have already mentioned that in adult motor behaviour continuing movements are merely subelements in movement sequences. The problems of the position of a continuing movement in space (i. e. its relationship to gravity) will be covered in detail in the next section. First of all, we will consider some fundamental aspects of speed, intensity of movement impulses, and the role played by the weights of parts of the body within the field of gravity.

Note
Assuming a movement impulse of a given intensity, we can make the following statements:

1. If the movement is *upward*, weights in the body have to be not only *moved* but *lifted;* this reduces the speed at which the movement can be performed.
2. If the movement is *horizontal,* the weights need only be *moved* and *prevented from falling;* this results in the least slowing down of the movement.
3. If the movement is *downward*, the weights have to be not only *moved* but *braked;* this speeds up the performance of the movement.
4. If the movement is horizontal and slightly downward, this tendency to speed up makes the movement impulse maximally effective. This is evident in an economical gait (see *Gangschulung*).

3.3.2 Buttressing Continuing Movement and Changing the Support Area

Continuing movement can be checked through buttressing or by changing the support area.

Definition: When a continuing movement is checked by a *counterweight,* a *counteractivity* or a *countermovement,* we call this *buttressing* the continuing movement.

Note
The movement which checks a continuing movement goes in the opposite direction to the continuing movement; the latter, being the initial, purposive movement, is called the *primary movement.* Defining the primary movement makes it easier to recognize buttressing when analysing movement.

Buttressing by Counterweighting

Buttressing a continuing movement by counterweighting is an automatic equilibrium reaction. The counterweight works against the horizontal component of the continuing movement, which has been defined as the primary movement. Because the horizontal component of the primary movement moves weight in the direction of movement, it has an accelerating effect upon the movement. Buttressing by counterweighting moves weight in the opposite direction, slowing the movement down.

Since a buttressing counterweight consists of inherently mobile body segments or parts of segments, we call it an 'activated passive buttress'. The term 'passive buttress' refers to the mass of the counterweight and 'activated' indicates that the lever of the counterweight has achieved the necessary length through muscular activity. Purely passive buttressing hardly ever occurs normally. The body segments available to act as buttressing counterweights represent, as it were, a *potential counterweight*. As soon as a counterweight is required, it is provided by a continuing movement. The musculature bridging the joints performs moving, lifting and fall-preventing work to ensure that the lever is of the required length and appropriately adjusted in space.

Definition: If the spatial path of the critical DP of a continuing movement is mainly horizontal and more or less straight, and there is no counterweighting by activated passive buttressing, the continuing movement is checked by *changing the support area*.

Note
Buttressing continuing movement by changing the support area is done in the direction of the primary continuing movement.

When the weights of primary movement and activated passive buttressing balance each other out, the distance points on the body segments constituting the counterweight go out in the opposite direction to the primary movement and the support area remains unchanged: this is a movement keeping a *constant location* (see Fig. 158). Constant-location movement also occurs when the weights are displaced vertically, as they only need to be lifted or prevented from falling.

If the horizontal component of a primary movement predominates slightly, the pressure exerted by the body upon its base changes, moving in the direction of the primary movement. If, at the same time, parts of the body leave the base support, the support area becomes smaller and moves slightly in the direction of the primary movement. The weight of the body parts thus lifted up functions as an activated passive buttress, even when distance points of those body parts move in the direction of the primary movement. This is a constant-location movement sequence with change in the area of pressure on the support area (see Fig. 159).

If the horizontal components of the primary movement predominate strongly the support area changes, moving in the direction of the primary movement, checking it. This is a movement which *changes location* (see Fig. 160).

In *location-changing* movement sequences, the economical form of activated passive buttressing is to relinquish areas of contact between the body and its base support. The body segments lifted up go into free play and hang from the body, thus becoming weights positioned on the braking side of the *bisecting plane* (see p. 143). The equilibrium reaction of activated passive buttressing is thus achieved without the braking weights moving in the opposite direction to the primary movement, because in being moved in the direction of the primary movement they lose contact with their base support and at once have a braking effect.

118

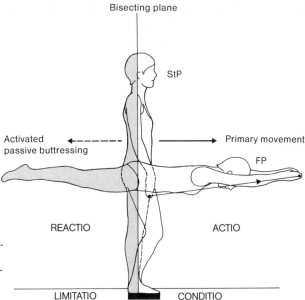

Bisecting plane

StP

Activated ← – – – – → Primary movement
passive buttressing

FP

REACTIO   ACTIO

Fig. 158. Constant-location
movement sequence from standing with the weight on the left
leg to the horizontal balance position. *StP,* starting position;
FP, final position

LIMITATIO   CONDITIO

Note

For therapy-oriented functional analysis of a location-changing sequence of movement, we need to know whether the new support area:

1. Is that of the final position of the sequence, or
2. Is only a temporary one in the course of a to-and-fro movement, or
3. Serves as the starting position of a new movement impulse in a different direction, or
4. Is that of a single phase in continuous locomotion.

Reference to the bisecting plane (p. 143) enables us to differentiate between the accelerating weights, belonging to the primary movement, and the braking weights,
belonging to the activated passive buttressing.

Normal gait, regarded for therapeutic purposes as a series of incomplete movements
which can be divided into individual step phases, is the representative example of
point 4. The low intensity of the economical activity in normal gait on a horizontal
base depends both upon the inertia of the rectilinear, space-consuming, forward movement of BS thorax and BS head and upon the slight predominance of weight in front
of the bisecting plane, the accelerating tendency of which provides a perpetual impulse
to movement (see *Gangschulung*).

- *Examples*

Starting position: Standing on the left leg. *Final position:* Horizontal balance (Fig. 158).
Constant-location movement sequence. The support area in the starting position is the

smallest area which includes the whole contact between the left sole of the foot and the floor. The left leg is supporting, the right leg is parked.

Critical DP: Tip of the third finger of the right hand.

Direction of movement: Forwards and very slightly upwards.

Spatial path of the critical DP: Straight line.

Critical fulcrum: Left hip joint.

Primary movement: Actio (see p. 144): Mixed continuing movement. The critical DP on the third finger of the right hand moves forward with flexion and external rotation at the right shoulder joint. DP acromion performs a co-rotational continuing movement cranially and medially, during which the wrist and the elbow are stabilized at zero. DP left iliac spine performs a counter-rotational continuing movement forwards and downwards flexionally at the left hip joint while the right hip joint is stabilized in extension. To relax the left ischiocrural musculature in order to allow flexion at the left hip joint, the left knee has moved slightly forwards and downwards.

Conditio: The pressure exerted by the sole of the left foot upon the ground must remain constant.

Activated passive buttressing in the final position: Since the bisecting plane meets the ground approximately between the forefoot and the heel of the left foot, the whole of the right leg and the part of the left leg which is behind the bisecting plane are involved in activated passive buttressing.

If the weight of the primary movement becomes too much, the left arm moves to a position in the mid frontal plane of the thorax. One could then say that the hand and part of the forearm are also part of the activated passive buttressing.

Activity states: The left leg is supporting, the remaining body segments in free play. BSs pelvis, thorax and head are aligned along the long axis of the body, which is almost horizontal. They are activated in fall-preventing extension as part of a continuing movement. At the left hip joint the fall-preventing component is made up by transverse abduction and extension.

Starting position: Standing. *Final position:* Standing on tip toe with flexion at the knee joints and extension at the hip joints (Fig. 159). Constant-location movement sequence with reduction of support area.

Critical DPs: Right and left greater trochanters.

Direction of movement: Forwards.

Spatial path of the critical DP: Straight line.

Critical fulcrum: Metatarsophalangeal joints of the large toes.

Primary movement: Co-rotational continuing movement. The right and left greater trochanters (flexion/extension axes of the hip joints) move forward with extension at the hip joints and flexion at the knee joints; DPs right and left lateral malleoli move forwards and upwards as plantar flexion occurs at the ankle joints.

In reaction, DPs right and left hands have moved backwards and slightly upwards, providing active passive buttressing.

Conditio: There should be no step forward.

Starting position: Sitting upright on a bench. The dorsal aspects of the thighs are the surfaces of contact between the body and its base support. *End position:* Like the start-

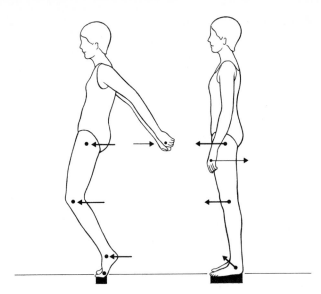

Fig. 159. Constant-location movement sequence with reduction of support area

ing position, reached via intermediate prone lying (Fig. 160). Location-changing movement sequence.

Conditio: The patient's plane of symmetry should remain vertical in the starting, intermediate prone, and end position, should correspond to the bisecting plane, and should remain parallel from one position to the next.

In the first phase of the movement we see that some of the distance points of the activated passive buttressing, e. g. the right and left feet, go in the opposite direction to the primary movement, while others, e. g. the right iliac spine, go in the same direction as the primary movement. Beneath the iliac spine, the right buttock loses contact with the base support and the pelvis hangs on the right-hand side from the thorax. Thus, the major part of the pelvis and both legs become braking weights so that the right hand, still in free play, can make contact with the edge of the bench without too great acceleration, go into supporting function, and rotate the body onto its stomach with the required braking activity performed by the elbow extensors. Push-off activity in the left hand initiates continued rotation of the body and the restoration of its long axis to the vertical. A high intensity of economical activity is evident in this movement sequence as the load-lifting and braking muscle activity required to deal successfully with the up-and-down components of the movement is considerable.

Buttressing by Counteractivity

Definition: Buttressing continuing movement by counteractivity is called *active buttressing*. The effect of a primary movement or of movement impulses is checked or halted by antagonistic muscle activity. This activity is a sophisticated equilibrium reaction, indispensable in movements requiring skill.

121

122

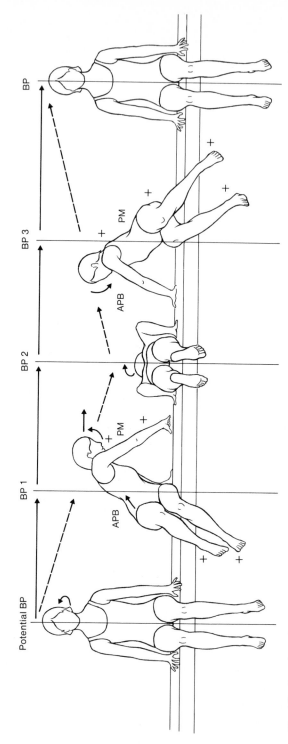

Fig. 160. Location-changing movement sequence from the starting position sitting upright on a treatment bench through prone lying and back to sitting upright. *BP*, bisecting plane; *APB*, activated passive buttressing; *PM* primary movement

Note

We speak of a sophisticated equilibrium reaction in contrast to the spontaneous equilibrium reaction of activated passive buttressing or changing the support area. To use the analytical concepts *actio* and *reactio* (see p. 144), the primary movement is the *actio* and the activated passive buttressing and changes in the support area are the *reactio*. Active buttressing is the limitation *(limitatio)* of a primary movement *(actio)* which fulfils particular conditions *(conditio)* and as such causes the specific details of a movement sequence to be performed (see *Ballgymnastik*, p. 3).

When continuing movement is checked by active buttressing, no further displacement of weights occurs. Unnecessary strain can thus be avoided or, if unavoidable, can be distributed.

● *Examples*

Active buttressing occurs continuously at the level of the thoracic spine through its dynamic stabilization in the neutral position during manual activities and is an expression of economical motor behaviour. Lack of active buttressing in the thoracic spine causes the lever of the upper extremity to become much longer, and signs of strain appear in the shoulder girdle and in the lumbar and cervical spine.

The accelerated primary movement of the hands downwards is actively buttressed by extensional activity in the thoracic spine which is stabilized at zero. When the hands move in the opposite direction, active buttressing again takes place in the thoracic spine, this time in the form of flexional activity of the abdominal musculature (Fig. 161).

Raising the right arm, with flexion and abduction at the shoulder joint, is a continuing movement running distal-to-proximal, with the critical fulcrum at the right sternoclavi-

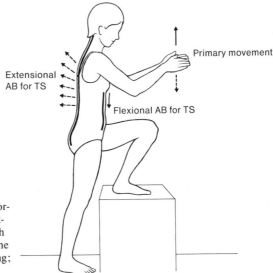

Primary movement

Extensional AB for TS

Flexional AB for TS

Fig. 161. Active buttressing in the thoracic spine during accelerated symmetrical movements of the hands through stabilization of the thoracic spine in the neutral position. *AB,* active buttressing; *TS,* thoracic spine

cular joint. It is actively buttressed at the level of the thoracic spine by muscle activity as if for negative rotation of the thorax with flexional and right concave lateral flexional activity in the thoracic spine. The active buttressing needs these three components because the movement is executed by one arm only (Fig. 162).

In discussing supporting function above (pp. 91 ff.) we have already indicated the role of the rotational components in providing active buttressing in this activity state. Active buttressing plays an important role in the treatment of limited joint mobility, above all in the reversal of muscular contractures. The procedure is as follows: just before the available movement tolerance is exhausted in a restricted movement excursion, active buttressing must be initiated and the unwanted continuing movement impulse at the next fulcrum stopped. The intensity of activity of the opposing forces can be increased within the body itself. It is possible, therefore, by engaging active buttressing, to make use of the full mobility tolerance of the various movement components available at the level of movement.

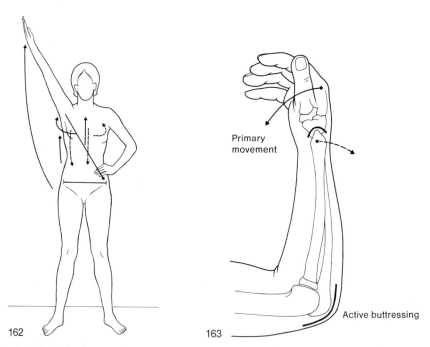

Fig. 162. Active buttressing at the level of the thoracic spine when raising the right arm by negative rotational activity in the thorax and flexional and right concave lateroflexional activity in the thoracic spine

Fig. 163. Primary movement: flexion at the right wrist; extensional active buttressing at the elbow joint

124

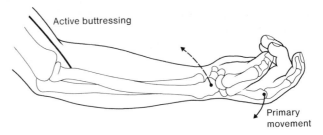

Fig. 164. Primary movement: extension at the right wrist; flexional active buttressing at the elbow joint

● *Examples*

Primary movement: Flexion of the right wrist (Fig. 163). With limited mobility, an un-wanted continuing movement would occur prematurely, in the shape of flexion at the elbow joint. This is prevented by extensional active buttressing at the elbow joint.

Primary movement: Extension of the right wrist (Fig. 164). With limited mobility, an un-wanted continuing movement would occur in the shape of extension at the elbow joint; this is prevented by active buttressing involving flexion of the elbow joint.

Remark: The quality of the active buttressing in these cases (Figs. 163, 164) is depen-dent upon the positional relationship of the hand movements to the rest of the body. If for instance, from a starting position with the shoulder joint at zero, the long axes of the forearm and the hand were placed sagittotransverse and the flexion/extension axis of the wrist frontosagittal, the active buttressing would consist of internal rotation when the wrist extends and of external rotation when it flexes. The flexion/extension axis of the wrist would then be parallel to the axis of rotation of the shoulder joint and the continuing movement would be transmitted to the shoulder joint by the rotational activity.

Primary movement: Flexion at the metacarpophalangeal joint of the index finger (Fig. 165). With limited mobility there would be a premature unwanted continuing movement in the shape of flexion of the wrist. This avoidance mechanism is prevented by extensional active buttressing at the wrist.

Primary movement: Extension at the metacarpophalangeal joint of the index finger with the distal joints stabilized in the neutral position (Fig. 166). To perform this move-ment forcefully and/or rapidly, the impulse to extensional movement at the metacar-pophalangeal joint of the index finger needs flexional active buttressing in the wrist.

Primary movement: Flexion at the distal joints of the index finger (Fig. 167). To achieve maximal range, the primary movement needs the resistance provided by extensional active buttressing at the metacarpophalangeal joint of the index finger.

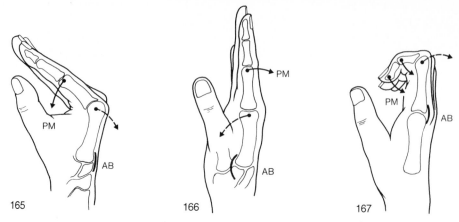

Fig. 165. Primary movement: flexion of the metacarpophalangeal joint of the index finger; extensional active buttressing at the wrist. *PM*, primary movement; *AB*, active buttressing

Fig. 166. Primary movement: extension of the metacarpophalangeal joint of the index finger; flexional active buttressing at the wrist

Fig. 167. Primary movement: flexion of the distal joints of the index finger; extensional active buttressing at the metacarpophalangeal joints

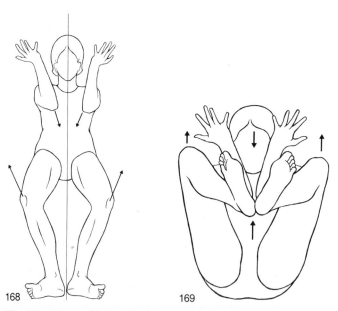

Fig. 168. Arms and legs move symmetrically

Fig. 169. Arms and head move in the opposite direction to the legs

Buttressing by Countermovement

Definition: The buttressing of continuing movement by a *countermovement* is a sophisticated equilibrium reaction. When two movement impulses occur in opposite directions, the movement tolerance of the switch points lying between can be fully utilized. This we call *buttressing movement*.

When the objective of the countermovement is exhaustion of the full tolerance of movement at particular fulcra, the problem which needs to be resolved is similar to that of active buttressing. Undesirable continuing movements must be cut out. In each of the opposing movements, the movement excursion which is to be completely exhausted must be in the line of fire of the continuing movement. Two movement impulses, one from distal, one from proximal, set respectively distal and proximal (or cranial and caudal) levers, pointers or gliding bodies in motion at the level of movement involved. In the vertebral column, the two movement impulses must invoke continuing effects in opposite directions.

- Such countermovements can be initiated at the extremities.
- The arms move symmetrically; the legs move symmetrically (Fig. 168).
- Head, arms and legs move in contrary directions (Fig. 169).
- One arm and the leg on the opposite side move in contrary directions (Fig. 170).
- One arm and the leg on the same side move in contrary directions (Fig. 171).

Activated buttressing by countermovement can also occur within the vertebral column, between two body segments, or within a body segment. The purest form of buttressing through countermovement is contrary movement of levers, pointers or gliding bodies at a particular fulcrum, which is discussed in the chapter on buttressing mobilization of the joints as a treatment technique (see p. 311).

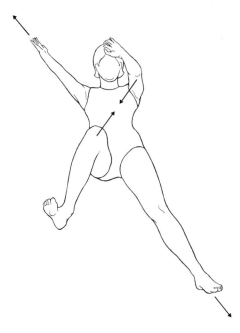

Fig. 170. The left arm and the right leg are moving in opposite directions; the right arm and the left leg are activated in their positions and work in active buttressing

- *Examples*

Starting position: Sitting upright on a stool (Fig. 172). The right arm describes a circular co-rotational continuing movement in the mid frontal plane. If the left arm performs a symmetrical continuing movement, the tendencies to lateral flexion in the thoracic and lumbar spine cancel each other out (Fig. 172a). At the level of movement of the shoulder girdle/thorax the movement tolerance at the critical fulcra, the right and left sternoclavicular joints, is used to the full. The movement performed by the right arm at the level of the thoracic and lumbar spine can alternatively be actively buttressed through right concave lateral flexional activity (Fig. 172b), or the head can join in, increasing the left concave lateral flexional activity by a continuing movement in the vertebral column (Fig. 172c). The lumbar spine is the critical fulcrum in this movement sequence. Active buttressing is provided by internal rotational activity of the pelvis at the right hip joint.

End position of the all-fours stance for mobilization of the vertebral column in extension (Fig. 173). The buttressing continuing movement of the left arm and the head answer the primary movement performed in the contrary direction by the right leg.

Note

All the forms of buttressing can occur in isolation or in combination. They have in common the tendency, in checking the primary movement, to keep constant the location of the body as a whole.

128

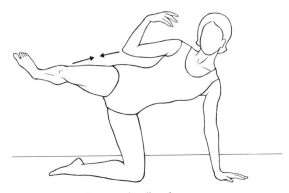

Fig. 171. One arm and the leg on the same side move in opposite directions

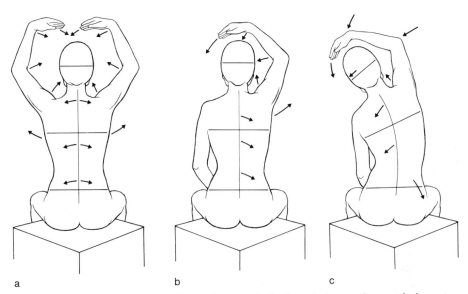

a   b   c

Fig. 172a–c. Continuing movement of the right arm in the frontal plane. **a** Symmetrical counter-movement; **b** active buttressing; **c** co-rotational movement of the head and the thoracic and lumbar spine

Activated passive buttressing keeps the body's location constant by shifting weight in the contrary direction to that of the primary movement, while active buttressing and buttressing movement halt the primary movement. Obviously, then, in economical motor behaviour, it must be made clear whether a movement sequence is planned to be of constant location or changing location, because only the change in the support area goes in the same direction as the primary movement. Active buttressing is therefore also the economical equilibrium reaction in locomotion. If the primary movement lacks

129

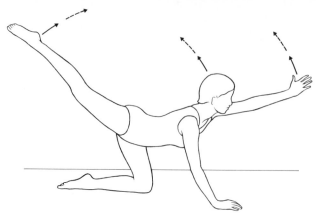

Fig. 173. Final position of all-fours exercise to mobilize the vertebral column in extension. The movement of the right leg is buttressed by the countermovements of the head and left arm

weight, the deficiency can be made up by increasing the intensity and speed of movement. However, increasing the speed increases the danger of falling, i. e. an unwanted change in the support area: forwards when the speed is increased too much, backwards when it is increased too little.

● *Example*
Starting position: Sitting upright on a stool. Complete exhaustion of the flexion and extension potential of the right elbow joint through buttressing movement and active buttressing (Fig. 174).
Elbow joint: Distal DP: styloid process of radius; proximal DP: head of humerus.
Shoulder joint: Distal DP: lateral epicondyle of humerus; proximal DP: inferior angle of scapula.
Sternoclavicular joint: Distal DP: acromion; proximal DP (in this case an axis): the frontotransverse diameter of thorax.

Flexion: The DPs of the elbow joints move closer together while the fulcrum moves out laterally to the right/cranially/dorsally (buttressing mobilization). The forearm goes into supination; the wrist is in slight extension and the fingers in slight flexion. The hand is caudal to the shoulder, which is in flexion/abduction/internal rotation, and the acromion has moved ventrally/medially. This movement of the shoulder girdle is actively buttressed in the thoracic spine by positive rotational activation of the cranial pointer, the frontotransverse diameter of thorax (Fig. 174a).

Extension: The distance between the DPs of the elbow joints increases while the fulcrum, moving medially/caudally/ventrally, takes up a position between them (buttressing mobilization). The forearm moves simultaneously into pronation and the wrist maintains slight extension and the fingers slight flexion. The hand is ventral/caudal to the shoulder joint, which is in extension/adduction/external rotation, and the acrom-

130

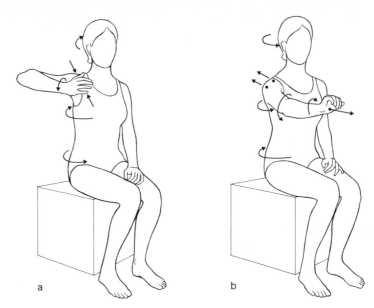

Fig. 174a, b. Complete exhaustion the of flexion and extension potential at the elbow joint through active buttressing and buttressing movement

ion has moved dorsally/laterally. This movement of the shoulder girdle is actively buttressed in the thoracic spine by negative rotational activation of the cranial pointer, the frontotransverse diameter of thorax (Fig. 174b).

The sequence of movement in this exercise is based upon a precise functional analytical concept. It is important that both movement components of the elbow joint are buttressed by the countermovements. Restricted flexion or extension at the elbow can only be overcome in self-training if the external rotation of the upper arm at the shoulder joint during extension of the elbow is buttressed by pronation of the forearm, and the internal rotation of the upper arm at the shoulder joint during flexion of the elbow is buttressed by supination of the forearm. Because the exercise is planned in every detail, the patient learns quickly and easily, particularly since the direction of movement of the critical distance points is a straight line.

Starting position: Sitting upright on a high bench to allow the long axis of the body to be vertical even if the flexion tolerance at the hip joints is limited (Fig. 175).
Planned movement sequence: Rising from the treatment bench to the standing position without the hands pushing off from the bench. The contact between the soles of the feet and the floor, which will be the support area at the end of the movement, is brought as far back as possible under the hip joints without the heels losing contact with the floor. The patient may need to wear shoes with heels. Since BSs pelvis, thorax and head, aligned in the long axis of the body, cannot be inclined forward by flexion at the hip joints in order to shift enough weight over the feet, the weight of the arms has to help, accelerating forwards.

131

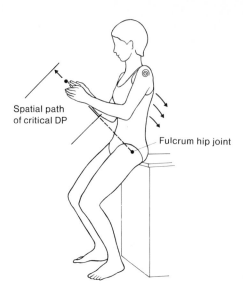

Spatial path
of critical DP

Fulcrum hip joint

Fig. 175. Rising from a sitting position by accelerating the weight of the arms

The critical distance points of the primary movement are the fingertips of both hands. The direction of movement is a straight line forwards/upwards, parallel to the path to be described by the right and left greater trochanters and the body segments over them. The sternoclavicular joints are the critical fulcra of the accelerated arm movement. Flexional stabilization of the thoracic spine in the neutral position provides active buttressing for the accelerated arm movement.

The distance covered by the accelerating arm movement is slight. Critical DPs fingertips travel 30–40 cm. The important features of the movement are: accelerating the arm movement, ending it before the elbow joints are fully extended, holding the final position, and waiting until the reaction of the accelerated weight of the arms carries the weights of BSs pelvis, thorax and head over the feet with little strain on the quadriceps.

3.3.3 Avoidance Mechanisms and Their Influence on the Patient's Contact with the Environment

Definition: Uneconomical, undesired continuing movements deviating from the direction of movement, or changes of support area, or buttressing of continuing movements, due to pain, stiffness, hypermobility, or lack of muscular strength or co-ordination, are called 'avoidance movements'.

This means that all the observation criteria established in sections 3.3.1 and 3.3.2 as fundamental for therapy-orientated analysis of movement sequences are at the same time criteria of avoidance movements. What is lacking in avoidance movements is economy and achievement of the purpose of the movement. Since an avoidance movement is as automatic as an equilibrium reaction, we use the term 'avoidance mechanism' or, when the avoidance mechanism relates specifically to gait, a 'limping mechanism'.

Causes
- Pain
- Restricted movement at joints
- Peripheral paralysis or paresis
- Disorders of the central motor apparatus
- Poor physical condition
- Bad habits

The physiotherapist's task
- To recognize an avoidance mechanism through observation.
- To assess the cause of an avoidance mechanism once the patient's functional status is established (see p. 213).
- To determine whether the avoidance mechanism is reversible. If it is, therapy must be planned to overcome it. If it is not, the physiotherapist must decide whether the existent avoidance mechanism is the best compromise possible, i.e. the body itself has found the best solution. If this is not the case, the physiotherapist will look for the most economical compromise and plan the therapy to achieve it.

Frequent forms of avoidance mechanism
1. Within the primary movement, part-weights are moved contrary to the main direction of movement (Fig. 176, 'spinning' exercise).

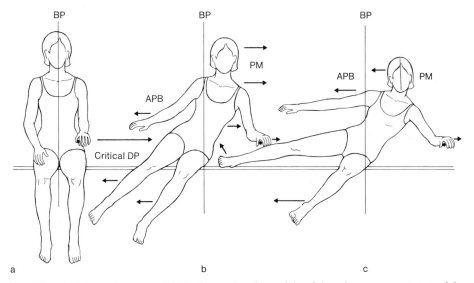

Fig. 176a–c. 'Spinning' exercise. Within the accelerating weight of the primary movement, partial weights are moved contrary to the direction of the primary movement. *BP*, bisecting plane; *PM*, primary movement; *APB*, activated passive buttressing

133

2. In a planned change of support area, the primary movement causes weight to be shifted in the direction of locomotion; activated passive buttressing at another level of movement makes this ineffective (Fig. 177, automatic reactive stepping).
3. If movement at a particular fulcrum is restricted, the fulcrum is bypassed within a continuing movement, which usually results in a deviation from the direction of movement: for instance, raising a hand when there is reduced extension at the elbow joint (Fig. 178). When the bypassed fulcrum is the critical fulcrum of a continuing movement, an unwanted continuing movement occurs, as in raising an arm when mobility at the shoulder joint is restricted (Fig. 179) or stretching it when extension at the elbow joint is limited (Fig. 180).
4. Muscular insufficiency may make it impossible to complete a continuing movement, and an undesired buttressing movement may take place instead (see Fig. 181, training of the abdominal musculature).
5. In bending of the long axis of the body forwards at the hip joints, the greater weight of BSs pelvis, thorax, head and arms prevents automatic extensional stabilization of the lumbar spine and the area of the lumbosacral junction (Figs. 182, 183).
6. Pathological patterns of movement due to disorders of the central nervous system are well known. They represent an accumulation of avoidance mechanisms which are treated by established methods (e. g. Bobath, Knott, Vojta). Functional analysis both reveals additional complications and suggests facilitations by means of therapy adapted to the patient's constitution, statics and mobility (see p. 144).

- *Examples*

The 'spinning' exercise from *Gangschulung:* Changing position from sitting on the right and left ischial tuberosities to sitting on the left greater trochanter (Fig. 176).

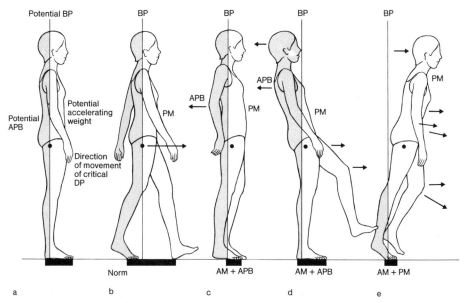

Fig. 177a–e. Limping mechanisms in automatic reactive stepping. *BP*, bisecting plane; *PM*, primary movement; *APB*, activated passive buttressing; *AM*, avoidance mechanism

Critical DP: The points of contact between the thumb and the second and third fingers of the left hand.
Direction of movement: To the left.
Spatial path of the critical DP: Straight line.

a) *Starting position:* Sitting upright on a treatment bench. BS legs: The lower legs are hanging in free play. The dorsal aspects of the thighs and the area around the tuberosities in BS pelvis are the areas of contact with the treatment bench. BSs pelvis, thorax and head are aligned in the long axis of the body, which is vertical. BSs pelvis and head are potentially mobile; in BS thorax, the thoracic spine is dynamically stabilized at zero. BS arms: The right arm is in parking function, the left arm in free play. The bisecting plane and the plane of symmetry are identical because the direction of movement is to the right.
b) Economical movement sequence: When the support area has moved slightly to the left and decreased (patient sitting on the left trochanter), the accelerating weight of the primary movement and the braking weight of activated passive buttressing balance each other out.
c) Avoidance mechanism: Within the accelerating weight of the primary movement, the head has inclined to the right, against the direction of movement, in order to reduce the accelerating weight. If the critical DP moves much further to the left, the body will fall on the left side.

'Automatic reactive stepping' as described in *Gangschulung* (Fig. 177).
Critical DP: The right greater trochanter.
Direction of movement: Forwards.
Spatial path of the critical DP: Straight line.

a) *Starting position:* Standing upright on both legs (Fig. 177 a).
b) Economical movement sequence: Reactive step taken by the left leg. The potential braking weights situated behind the bisecting plane do not effect activated passive buttressing as long as the right leg is in contact with the base support; only the left arm swinging backwards functions as activated passive buttressing (Fig. 177 b).
c) The greater trochanter is brought forwards in the primary movement and the weight is taken off the left leg, but the shoulder girdle with both arms, is moved backwards and effects activated passive buttressing. Since what was intended was a step forwards, this is an avoidance mechanism in the form of unwanted activated passive buttressing (Fig. 177 c).
d) The greater trochanter is brought forwards in the course of the primary movement and the left leg, freed from supporting weight, is actively raised in the primary movement, but the shoulder girdle, parts of the thorax and the head move reactively contrary to the direction of movement, providing activated passive buttressing to the leg raised. This is an avoidance mechanism, for a location-changing movement sequence was planned, whereas a constant-location sequence has been performed (Fig. 177 d).
e) The greater trochanter and the body segments over it have been moved too far forwards, the step has not yet taken place, because of an avoidance mechanism in the form of extensional active buttressing at the knee against the direction of move-

ment. As the accelerating weight of the primary movement is too great, the path of the critical DP deviates downwards. The consequence is another avoidance mechanism in the form of a lunge step forwards by the left leg. The step is uneconomical because the excessive acceleration calls forth braking activity against the direction of movement (Fig. 177e).

Raising the hand with limited extension at the elbow joint (Fig. 178).
Starting position: Standing upright with right shoulder and elbow in slight flexion; the long axis of the forearm is sagittotransverse.
Critical DP: Radial aspect of the right wrist with the flexion/extension axis vertical.
Direction of movement: Upwards.
Spatial path of the critical DP: Straight line. Because extension at the right elbow joint is restricted, the movement instruction cannot be carried out. An avoidance mechanism causes the movement to deviate backwards out of its intended direction and the spatial path changes from straight to circular.

Raising the arm with restricted function at the shoulder joint (Fig. 179). If flexion at the right shoulder joint is restricted, an avoidance mechanism in the form of an unwanted continuing movement comes into play, consisting of extension of the thoracic and lumbar spine. The long axis of the thorax inclines backwards.

Avoidance mechanism elicited by restricted extension in the right elbow but with the direction of movement maintained (Fig. 180).

Insufficiency of the abdominal musculature when attempting to raise the legs in extension (Fig. 181).

Fig. 178. Deflection of the direction of movement due to restricted extension at the elbow joint

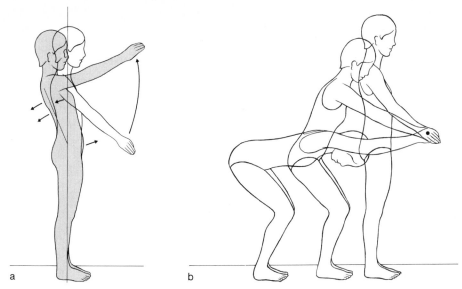

Fig. 179 a, b. Raising the arm with restricted function at the shoulder joint. **a** Functional insufficiency: flexion at the right shoulder joint. Avoidance mechanism: starting position is the upright standing posture. The usual forward/upward direction of movement of the right hand is achieved, but only because, in a continuing movement, the lack of flexion at the shoulder joint causes the hips to be displaced forwards and the shoulders backwards. The desired flexion occurs only partially or not at all. **b** The wrist is defined as a fixed point in space. The movement is redefined as backward/downward. Instead of the hand moving upwards/forwards, the shoulder moves backwards/downwards. The patient is necessarily forced to step backwards. The flexion at the shoulder joint takes place

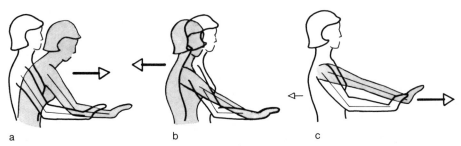

Fig. 180 a-c. Avoidance mechanism when extension is restricted at the right elbow joint. Starting position: standing upright. **a** In the attempt to extend the elbow, the right hand does move forwards (orientation outwards from the body), but the elbow remains flexed. The avoidance mechanism is an unwanted continuing movement: DP right shoulder, the thorax and the head move forwards with the hand. **b** Distal to the affected elbow, the wrist is defined as a fixed point in space. The direction of movement is redefined as backward: DP right shoulder, the thorax and the head move backwards (orientation outwards from the body). DP right wrist must not change its position in space. Result: the elbow is extended. **c** In normal motor behaviour and with the elbow freely movable, extension of the right elbow from the starting position shown takes place in the following manner: the primary movement is forwards. As the elbows is to be extended, the movement is limited by automatic active buttressing, the thoracic spine being extensionally activated in the neutral position and the right shoulder blade fixed in retraction to the vertebral column

137

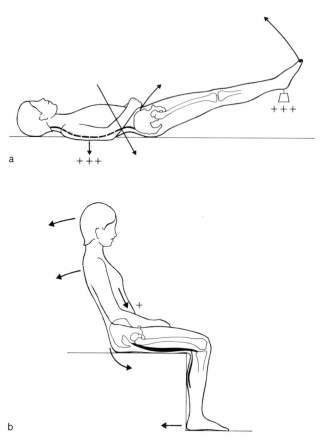

Fig. 181. a Typical avoidance mechanism when attempting to raise the legs in extension in the supine position. **b** Correction of the mechanism by hanging the body from the seat by the flexors of the knee joints

Starting position: Lying supine with arms parked on the floor close to the body.
Critical DP: Tips of the toes.
Intended critical fulcrum: The hip joints.
A co-rotational continuing movement arises, extensional at the joints of the toes and knees and flexional at the hip joints. BS legs goes into free play and is suspended by the flexor muscles from BS pelvis. The spine in this segment (the lumbar), unprepared for the strain imposed upon it, is deformed extensionally, so that the pelvis, as the proximal level, performs a buttressing movement to the primary movement of the legs through flexion at the hip joints. This avoidance mechanism can only be prevented if the abdominal musculature is strong enough to achieve flexional stabilization of the lumbar spine at zero and the thoracic and cervical spine is simultaneously stabilized at zero. These stabilizations does not occur automatically and cannot take place at all if the weight of BS legs is too great in relation to the rest of the body, in which case nothing will be achieved even by practice.

138

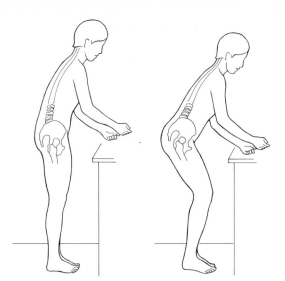

Fig. 182. When the long axis of the body inclines forwards, lumbosacral extensional anchorage does not take place

By altering the starting position the avoidance mechanism can be prevented: the new starting position is sitting upright, preferably across the corner of a stable piece of equipment. BSs pelvis, thorax and head, stabilized in the long axis of the body, are inclined backwards extensionally at the hip joints. The arms are parked on the thighs or crossed on the chest. With the aid of the flexors at the knee joints, the patient suspends himself from the seat. The strain upon the flexor components of the ischiocrural musculature at the knee joints stimulates extension of the pelvis at the hip joints. This movement of the pelvis at the hip joints makes the avoidance mechanism of extension in the lumbar spine impossible. The fact that the body is suspended from the seat means that the excessive weight of the legs no longer matters. The *conditio* that the distance between the navel and the xiphoid process must not change keeps the thoracic spine at zero and causes the oblique and transverse abdominal musculature to be activated in addition to the rectus abdominis muscles.

Figure 181 shows both the avoidance mechanism, caused by unwanted buttressing movement at the critical fulcrum the hip joint (flexion at the hip joint initiated by the proximal lever – the pelvis – and extension of the lumbar spine), and its correction.

Figure 182 illustrates the absence of lumbosacral muscular anchorage which occurs for example, during work in the kitchen. A therapeutic exercise for learning spontaneous extensory stabilization of the lumbar spine with lumbosacral anchorage is the ball exercise depicted in Fig. 183.

3.3.4 Summary

We are now in a position to carry out movement analysis, which forms the foundation of a problem-centred movement therapy. Let us first recapitulate the routine followed in observing movement:

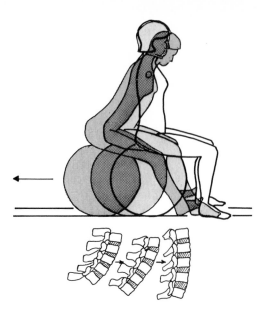

Fig. 183. Exercise with a gymnastic ball to learn spontaneous extensional stabilization of the lumbar spine with lumbosacral anchorage

1. During a movement we register the changes which occur in the contact between the patient and his environment, i.e. the base support, supportive device or suspension device. Thus we recognize changes in the support area.
2. We define the direction of movement and look for the critical distance point.
3. With the aid of the bisecting plane (see p. 143) we assess the displacement of weight as the sum of accelerating and braking weights, and differentiate between horizontal and vertical displacements of weight.
4. We note the changes in position at the joints and assign them either to the primary movement and the change of support area or to the buttressing activities.
5. Since we are now familiar with the activity states and know that the activation of the musculature is governed by the weights which have to be lifted, held, moved, braked or prevented from falling, we can reconstruct the activities which must have called forth the movements.

3.4 Observer's Planes

When the physiotherapist observes the posture and movement of a patient, she must compare her optical perceptions with hypothetical norms. In doing this, she must beware of being misled by perspective. However, misinterpretation can often be avoided if we make use of 'observer's planes' (Fig. 184).

The observer's eyes must be horizontal and their plane of symmetry must be vertical. The patient under observation should stand at a reasonable distance from the observer. In the description of observation, the following points should be borne in mind:

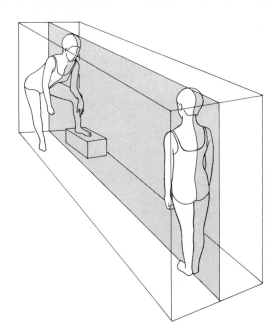

Fig. 184. 'Observer's planes' to aid in accurate observation of the patient: horizontal, parallel and bisecting planes

In front: The patient is in the observer's field of vision, directly in front of but at some distance from her. 'Behind' is not used. *Above* is the distance from the floor; *below* is the floor on which the patient is lying, standing, sitting or walking. *Right* and *left* are used by the observer in relation to her own right and left. If she can see the patient's left side, then the patient when walking forwards is moving left for the observer. If she is watching from behind and the patient walks forwards, he is, for the observer, moving from nearer and in front to further away and in front. If the distance between the patient and the observer increases, seeing the patient in perspective makes him appear smaller in the observer's eyes. The experienced eye of the observer corrects this diminishing image automatically back to the patient's actual size. Despite this ability of the observing eye to correct the distortion due to perspective back into the reality of experience, however, the following observer's planes should be firmly grasped.

3.4.1 Observer's Horizontal Plane

This plane is horizontal in space. It is the horizontal transverse plane through the eyes of the observer, projected forwards until it meets the patient, dividing him into an upper and lower part. The physiotherapist needs to adjust her horizontal plane to a level which is appropriate to the patient's height.
In observation of movement, the most extensive displacement in relation to the horizontal base takes place in the observer's horizontal plane.

3.4.2 Observer's Parallel Plane

This plane is vertical in space. It is the vertical frontal plane through the eyes of the observer, displaced forwards in a parallel movement until it meets the patient. It enables one to judge actual distances on the patient's body and make comparisons between them. For this, it is necessary that the demarcating distance points have been brought into the parallel plane, approximately level with the observer's eyes. There should, of course, be no switch points of movement between the critical distance points which might alter the distance. If that is impossible, the observer needs to keep a careful eye on these switch points.

● *Examples*

When comparing the length of the arms or legs, it is important to ensure that the positions of the elbow and the knee joints are symmetrical. The distance points selected should be on the forearm and lower leg and on the upper arm (or, as a compromise, on the acromion) and thigh.

The frontotransverse diameter of the thorax should be compared with the distance between the right and left iliac spines. If a scoliosis of the vertebral column has forced these axes out of parallel, one should try to bring both axes simultaneously into the parallel plane; it is far more difficult to move them separately into the parallel plane and then relate them to each other from memory. In addition, we must notice the position of the ribs during inspiration and expiration.

Let us say we wish to assess the distance between a patient's shoulder joints: we can see its real size when we align DPs right and left shoulder joints to the observer's parallel and, if possible, horizontal planes. This enables us to judge whether the shoulder joints are an equal distance from the manubrium sterni. If we bring the two shoulder joints into the observer's bisecting plane, level with her eyes (see p. 143), the distance is zero according to her optical perception. If we move the connecting line between the shoulder joints in the observer's horizontal plane to an angle of 60° to the parallel plane, the optical distance perceived is half the actual distance as seen in the parallel plane (Fig. 185).

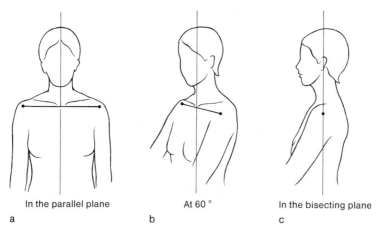

| In the parallel plane | At 60° | In the bisecting plane |
| a | b | c |

Fig. 185a–c. Distance between the right and left shoulder joints

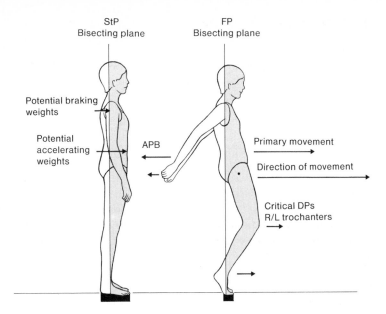

Potential braking
weights

Potential
accelerating
weights

APB

Primary movement

Direction of movement

Critical DPs
R/L trochanters

Fig. 186

Remark: Knowing these simple rules of perspective can be of great assistance to the physiotherapist and prevent errors. Her ability to correct the distortion of perspective enables her to adjust her observations continuously without constant recourse to measuring instruments during treatment of the patient.

3.4.3 Observer's Bisecting Plane

This plane is vertical in space. It is the vertical plane of symmetry of the eyes of the observer, projected until it meets the patient, bisecting him into a right and a left part. If the observer aligns her bisecting plane with the patient's centre of gravity, she can assess the potential accelerating and braking weights of the patient's body (see p. 117) in movement in a direction at right angles to the bisecting plane (Fig. 186; see also *Ballgymnastik*, p. 74).

Remark: The patient's centre of gravity is not visible, but it can be guessed at: it will be roughly above the place where the body exerts the greatest pressure upon its base support, if there is only one point of contact between the patient's body and its support. If there are several points of contact, ascertaining the position of the centre of gravity is more difficult, because it is usually above a place which has no contact with the base support (see p. 89). However, once the observer has become sensitized to perceiving in this way she can pinpoint the patient's centre of gravity fairly precisely in relation to the horizontal plane.

In observation of a movement sequence, the horizontal component of the direction of movement must be at right angles to the bisecting plane. As the movement sequence is carried out, the bisecting plane is displaced in the direction of movement and permits recognition of the accelerating and braking weights in every phase of the movement (see Fig. 176).

4 Instruction

It has already been shown, with many examples, how, in addition to the functional analysis of a movement sequence, expressed in 'therapist language', we need a formula for the instructions to get a patient's therapeutic exercises under way. This can best be achieved through the medium of 'patient language'. The words, gestures and actual physical handling of the patient which make up this language are simple and comprehensible; they are easy to follow because of the way they speak to the patient's kinaesthetic, tactile, auditory and visual perception.

It is common practice to differentiate between therapeutic exercise and habitual or natural movement. In physiotherapy this kind of contrast is rather arbitrary, because all physical exercise serves a therapeutic aim. If we instead make our differentiations according to the direction components of the primary movement, a system emerges which enables us to analyse all patterns of movement according to a single concept. We differentiate between:

1. Primary movements – called *actio* – determined by a single critical distance point which, because of the direction of movement prescribed for it, essentially decides the character of the movement pattern. When its direction component is in a straight line, horizontal and continuous, provided certain conditions named *conditio* are observed, the most economical form of locomotion is brought about. In the continuous change of support area we recognize the dominant equilibrium reaction, called *reactio*. This is called a location-changing sequence of movement.

   • *Example*
   Normal gait.

2. Primary movements called *actio,* again characterized by a single critical distance point, the direction components of which are in a straight line and horizontal. The *conditio* here, however, is that the support area must not be relinquished, though changes may if necessary occur within it. The dominant *reactio* will be counterweighting, in the form of activated passive buttressing. In a variant of this, the movement of the critical distance point is in a straight line and horizontal but goes alternately to and fro. These are location-constant sequences of movement.

   • *Examples*
   Horizontal Balance Position (see Fig. 158). In this exercise the contact between the sole of the foot and the floor constitutes the support area. The pressure exerted by the body upon it must remain constant throughout the movement sequence.

'Spinning' Exercise (see pp. 134–135). In the course of this exercise the support area diminishes and is slightly displaced in the direction of movement.

Reactive Arm Swing (see Fig. 306, p. 282). In the course of this exercise the critical distance points of the primary movement, the right and left greater trochanters, move alternately backwards and forwards while the support area diminishes in the direction of the primary movement each time the end position is reached.

In the primary movements described above we ascertained a dominant *reactio*. In exercises of this nature it is possible to achieve the learning target by making use of the body's automatic reactions. These exercises are prototypes of natural patterns of movement. Whether they are location-constant or location-changing depends upon the prevailing *conditio* which either limits the movement sequence, by forbidding changes to the support area, or causes it to continue, by suppressing activated passive buttressing. The effect of the *conditio* is called the *limitatio*.

3. Primary movements called *actio,* again characterized by a single critical distance point, whose direction components are in a straight line but vertical. If the direction of movement is upwards, weights have to be lifted; if it is downwards, the weights have to be braked and lowered. This exercise is location-constant. The *conditio* aims at a *limitatio* which ensures economical loading of the passive and active structures of the motor system. There is no dominant *reactio*. If the vertical direction component is to be maintained, every horizontal weight displacement needs to occur in two opposite directions at once, otherwise the support area will be changed. The potential for compensating horizontal weight displacement through active buttressing is very limited.

- *Example*

Horizontal and vertical type bending behaviour (see p. 279).

4. *Actios* realized by several simultaneous primary movements with a corresponding multiplicity of critical distance points. Their horizontal direction components cancel each other out, leaving as a resultant only a vertical component, that involved in lifting and lowering weights. The movement sequence is location-constant. During lifting the support area may diminish concentrically; during lowering it may increase concentrically. There is no dominant *reactio*. The important factor in this instance is the *conditio,* which, resulting in the *limitatio,* guides the movement sequence along the desired path to economy of movement and attainment of the learning goal.

- *Example*

The Classic Frog (see *Therapeutic Exercises*).

If, in the course of such exercises, the horizontal direction components of the simultaneously occurring primary movements do not balance each other out, the *reactio* which arises indicates the predominant horizontal direction component through the ensuing equilibrium reactions. This kind of exercise demands relatively great conscious control, i.e. *actio* with a good deal of *conditio*, since without this guidance it could not succeed. These exercises are, after all, carefully thought out movement sequences or changes of activity states. They mark out and circumscribe the patient's functional deficit in such a way that evasive movements can be avoided and the desired function, initiated by an unambiguous stimulus, must take place. The learning process by which this eventually becomes automatic is quite long.

Outline and Explanation of Movement Analysis and Treatment Programmes

Name of the exercise: An exercise which has proved successful should have a name. This could refer to the functional problem which the exercise is meant to resolve. However, it has become apparent that such names are not remembered by the physiotherapist and even less so by the patient. Other names spring spontaneously from the imagination; often it is the patient who gives an exercise its name. These invented names are easily remembered even if there is no obvious connection with the exercise itself.

● *Example*
'Change of position from sitting upright to sitting on right trochanter' would be a suitable functional name for one exercise; its invented name is the 'Spinning Exercise'. The 'spinning' refers to the movement of the critical distance points of the primary movement. The points of contact between the thumb and the tips of the second and third finger of the right hand move carefully to the right, and only to the right, as if forming a spun thread from a spindle; this is the *actio* of the movement instruction given to the patient.

Learning goal: The functional problem defined in the patient's function status (see p. 286) is to be partly or completely resolved with the aid of a movement exercise.

● *Example*
Hans S., 1944. Diagnosis: Chronic lumbago due to faulty posture of the vertebral column.

Functional problem: Pelvis raised on the right with thorax sliding downwards on the left; excessive flexion of the pelvis at the hip joints together with extra weight at the abdomen (obesity) causes strain upon the passive structures of the lumbar spine and reactive hypertonia of the lumbar musculature. In addition, a genu recurvatum has caused a loss of potential mobility of the pelvis at the hip and lumbovertebral joints. The poor statics of the leg axes disturbs the economical transfer of locomotor activity from the legs to the lumbar spine.

We can derive a number of learning goals from these functional problems. The patient must relearn:

- In walking, how to achieve correct functional load-bearing on the left leg in the roll-on phase in walking
- In standing, how to distribute his body weight between the right and left leg so as to correct the tilt of the pelvis, thus re-establishing the potential mobility of the pelvis at the hip and lumbovertebral joints
- In sitting and standing, how to align the pelvis, thorax and head in the vertical long axis of the body, making the reactive lumbar hypertonus superfluous

146

– How to move the pelvis at the left hip joint in all the movement components as near-ly lift-free as possible, with the thoracic spine dynamically stabilized at zero

Teaching method: An exercise which either partly or completely achieves the learning goal must become part of – be incorporated into – the patient's motor behaviour.

4.1 Functional Analysis in Therapist Language

Functional analysis is the physiotherapist's confrontation with the problem of how to teach movement and thus cure a patient. Through it, she is able to recognize the many functional elements in a sequence of movement, something of great importance when instructing the patient. To the extent that she recognizes, differentiates between and classifies these functional elements, she gains specific information which enables her to manage the learning steps individually.

4.1.1 Conception of the Exercise

By 'conceiving' an exercise we really mean inventing it.

1. If a movement sequence which occurs habitually in everyday life is disrupted by an avoidance mechanism, the physiotherapist must identify the faulty detail and plan an exercise based upon that detail. This exercise must make it possible for the pa-tient to practise the deficient function under simplified conditions; this is the only way in which an avoidance mechanism can be eliminated.

   - *Example*
   Malfunction in walking: The roll-on over the functional long axis of the left foot supporting phase is faulty (see p. 265). As a simplifiction of the normal walking pat-tern, the roll-on stage over the functional long axes is practised with both feet, sym-metrically and simultaneously. The step which would follow on is not performed; the exercise is location-constant. The rolling should be performed alternately for-wards and backwards. The learning goal is transferred into the area of the body's au-tomatic reactions, in that the primary movement demanded in the exercise is a sym-metrical, parallel pendulum swinging of both arms.

2. If the thoracic spine is no longer able to fulfil its weight-bearing function in the up-right stance, due to poor condition of the back musculature, the physiotherapist must invent an exercise which increases the lifting capacity of the thoracic spine without exposing the patient's circulation to excessive stress.

   - *Example*
   In the Classic All-Fours, the long axis of the body is horizontal with the back facing upwards. Co-ordination of the movements of the head, right arm and left leg, which taken place simultaneously, demands careful balancing of the weights of these ex-

tremities. These co-ordinated primary movements, because of the continuing effects of the direction components of their critical distance points, bring the weights of BSs pelvis and thorax within the area in which the thoracic spine is responsible for lift; the thoracic spine is in extensional and rotational stabilization and under extensional and rotational strain.

4.1.2 Position and Activation in the Starting Position

The conception of the movement sequence determines the starting position. In addition to lying, standing and sitting there are many other positions, such as occur in ball gymnastics. However, it is important for the analysis that the starting position impose a particular behaviour in regard to dealing with the body's weights.

Position in Space of the Critical Axes and Points of Contact with the Environment

In the starting position, the position of the critical axes must be registered in relation to gravity.

- *Examples*

Sitting upright on a stool about 45 cm high. BSs pelvis, thorax and head are aligned in the long axis of the body, which is vertical. The soles of the feet are in contact with the floor; they are the width of the hip joints apart and their functional long axes point forwards. The knee and the hip joints are in 90° flexion. The palms of the hands rest on the ventral aspect of the thighs.

In the starting position of the Classic Frog (see *Therapeutic Exercises*), the movement components of the joints of the extremities from proximal to distal must be explained to the patient precisely. They go to the limit of movement tolerance (end-stop) and form the starting position for the antagonistic movements as in PNF patterns.

Movement Components in Relation to the Neutral Position of the Joints

In the starting position, a note is made of movement components which deviate from the neutral position of the joints.

Movement Components at the Critical Joints in Relation to the Intended Primary Movement

According to what the direction components of the intended primary movements are, we can say which joints have a great tolerance of movement and which not.

- *Examples*

In the starting position of the Classic Frog the extremities have a great tolerance of movement all the way to the end position of the antagonistic pattern. In addition, the vertebral column has movement tolerance for flexion, lateral flexion and rotation.

If, from a standing position, critical DPs right and left greater trochanters are to initiate a primary movement forwards, the rather small movement tolerance for extension at the hip joints and the ample movement tolerance for flexion at the knee joints and extension at the proximal joints of the large toes allow one to foresee that the continuing effect of the primary movement will be that of a rolling-on along the long axes of the feet.

Distribution of Body Weights on a Base Support or Suspension Device, Against a Supportive Device and over a Support Area, and the Resultant Activity States of the Musculature

The arrangement in space of body segments and their contact with the environment provide information regarding the activity states of the musculature which holds together and stabilizes the inherently mobile system of the body.

● *Examples*
Sitting upright on a treatment bench. The region around the right and left ischial tuberosities and the dorsal aspects of the thighs form the area of contact between the body and its base support. If the long axis of the body is inclined forward, the dorsal aspects of the thighs and their base support will become part of the body's support area. The lower legs, suspended from the knee joints, are in free play, while the thighs are parked on the bench. The pelvis is potentially mobile at the hip and lumbar vertebral joints, the thoracic spine is dynamically stabilized at zero, and the head is potentially mobile at the cervical spine and the atlanto-occipital and atlanto-axial joints. The shoulder girdle is parked upon the thorax, and the arms are parked on the ventral aspects of the thighs, via the contact area of the palms of the hands.

In the Mobilizing All-Fours stance in extension (see Fig. 130), the left knee and the palm of the right hand form the areas of contact between the body and its base support. The support area is the smallest area which includes these two areas of contact. The right arm and the left thigh are in supporting function; they are the pillars of a bridge construction. BSs pelvis and thorax form the arch of the bridge; they hang suspended from the pillars and their downward-facing aspects are engaged in bridging activity. The head hangs from the musculature on the upper aspect of the bridge.

Intensity of Muscle Activity Required in Economical Activity and in Respiration

● *Examples*
In sitting upright on a stool the intensity of economical activity is minimal and respiration is at the normal resting rate.

In the starting position of the Classic All-Fours the intensity of activity is slightly increased but respiration remains the same as before.

In the activated starting position of the Classic Frog the intensity of the economical activity is high and the activation elicits inspiration.

149

Potentially Accelerating or Braking Weights in Relation to the Bisecting Plane of an Intended Primary Movement

This criterion only applies when the direction component of the critical distance point of the primary movement is predominantly horizontal.

● *Examples*

When standing upright in the starting position for the symmetrical roll-on over the long axes of both feet, the vertical frontal plane down the middle of the patient (seen from the side) becomes the potential bisecting plane for the movement sequence. The weights in front of the bisecting plane will accelerate the roll-on forwards and those behind will brake it, and vice versa for the roll-on backwards.

In the starting position of the Spinning Exercise (see p. 135), the plane of symmetry, which is vertical, will be the bisecting plane of the intended movement sequence. When the primary movement is to the right, the potentially accelerating weights are situated to the right of the bisecting plane and the braking weights to the left, and vice versa for a primary movement to the left.

4.1.3 *Actio – Reactio* of the Movement Sequence

The *actio* is that part of the movement sequence which is performed consciously. An *actio* can take place in many different ways. If the critical distance point has a predominantly horizontal direction component, equilibrium reactions come into play automatically: a *reactio* occurs as buttressing through counterweighting or a change of support area. When the critical distance point has only a vertical direction component, clearly visible equilibrium reactions of this kind do not occur.

Actio in the Form of the Primary Movement – Reactio in the Form of Activated Passive Buttressing – Reactio in the Form of a Change in Support Area

It is necessary, for the primary movement – the *actio* – that the physiotherapist can find the critical distance point or points whose spatial path most unambiguously adheres to the direction of movement, without deviation. The primary movement then becomes a true continuing movement. The critical distance point is guided in the prescribed direction, and the physiotherapist's task is to recognize the movement excursions at the neighbouring joints which participate in this prescribed movement, and if necessary, to analyse the activities which produce such movements.

If the direction of movement is predominantly horizontal, the *reactio* must now be analysed. We observe whether the distance points of the activated passive buttressing move contrary to the direction of movement or whether the counterweighting is provided by body segments or parts of them being lifted up from the support area. If the latter, we have a *reactio* in the form of a change in the support area. In most cases both forms of *reactio,* as just described, occur simultaneously. The physiotherapist must decide whether this is desirable.

Actio in the Form of Accelerating Weights – _Reactio_ in the Form of Braking Weights

In movement sequences where the direction component of the critical distance point of the primary movement is predominantly horizontal, it is worth comparing the accelerating weights of the _actio_ with the braking weights of the _reactio_. The bisecting plane, which may remain stationary or move with the direction of movement, permits a precise analytical differentiation between the form of _reactio_ defined as activated passive buttressing and the _reactio_ which, in the absence of a counterweight, brings about a change in support area in order to prevent a fall in the direction of movement.

- _Examples_
The step in normal walking.

If the patient is supine on the floor and the direction of the primary movement consists of a vertical and a horizontal component with the latter going from distal to proximal, the interplay between the accelerating primary weight and the braking counterweights can become very confusing. If the weights of the primary movement, which are to be lifted, are too heavy, they turn into braking weights, and the counterweights losing contact with the floor in reaction, have an accelerating effect. This 'topsy-turvy' state of affairs is only resolved when the longitudinal dimension of the primary weight is almost vertical.

Starting position: Lying supine. The joints of BSs pelvis, thorax, head and arms are at zero. The long axes of the upper arms are parallel to the long axis of the body; the elbows are fully flexed, and the flexion/extension axes of the wrists are frontotransverse and above and slightly caudal to the acromion. The palms of the hands are vertical and face towards the feet. The tips of the second to fourth fingers on both hands are the critical distance points of the primary movement. They move upwards and as far caudally as possible at an angle of about 45°. As soon as BSs pelvis, thorax and head, under the continuing effect of the primary movement, have been raised from their base by flexional activity at the hip joints, they and the arms become braking weights in relation to the direction of movement and the bisecting plane, which passes through the hip joints. The legs become accelerating weights as soon as they have been raised from their base by flexional activity at the hip joints. If their weight is insufficient to maintain equilibrium, the exercise cannot succeed. As soon as the critical distance point crosses the bisecting plane, becoming part of the accelerating weight, and BSs pelvis, thorax and head, aligned in the long axis of the body, have almost gained the vertical, the legs come back to rest on the base support. Their weight is no longer required to keep the critical distance point moving in the direction prescribed.

4.1.4 _Conditio – Limitatio_ of the Movement Sequence

In order to be able to formulate a _conditio,_ the physiotherapist must know the possible variants in which the _actio_ can be performed. To direct the _actio_ towards the variant required for therapy, the other variants must be ruled out by the _conditio_. This usually requires more than one *conditio*. The process of ruling out other possible variants is called *limitatio,* and it is only this that guides the movement sequence into the economical form prescribed by therapy. Each *actio* requires a limiting *conditio,* regardless of horizontal or vertical direction components of the critical distance points of the primary movements. If the direction of movement is predominantly horizontal, the *conditio* governs the *reactio;* if predominantly vertical, it serves to influence the distribution of work among the passive and active structures of the motor apparatus in the most economical way.

Conditio of Keeping a Constant Distance Between Distance Points on the Body – *Limitatio* in the Form of Stabilization and Active Buttressing

● *Examples*

Starting position: Sitting upright on a stool about 45 cm high. The *actio* demands that critical DP vertex is moved forwards and downwards by flexion at the hip joints. The *conditio* demands that the distances between DPs navel and xiphoid process and between DPs suprasternal notch and point of chin remain constant. The effect of the *limitatio* is to stabilize the thoracic spine at zero, enabling the head to remain in alignment in the long axis of the body wth the help of active buttressing in the form of dorsal translation of the head.

Starting position: Standing upright on the left leg. The *actio* demands that DPs right lateral and right medial malleolus move forwards and upwards. The *conditio* is that the distance between DPs right lateral malleolus and right greater trochanter remains constant. The effect of the *limitatio* is to stabilize the right knee joint at zero – i. e. extensional active buttressing of the knee joint. A further *conditio* demands that the distance between the DPs symphysis pubis and navel remains constant. The effect of the *limitatio* is to bring about active buttressing flexionally in the left hip joint and extensionally in the lumbar spine.

Conditio of Absolute and/or Relative Fixed Points in Space – *Limitatio* Through Limitation of the Primary Movement by Activated Passive Buttressing or Changes to the Support Area

● *Examples*

In the first example above, starting from sitting on a stool about 45 cm high, the *actio* was inclining the long axis of the body forward with flexion at the hip joints and stabilization of the thoracic spine at zero, while the *conditio* was that the head should remain in alignment with the long axis of the body. We now add a further *conditio:* that the contact between the soles of the feet and the floor should be a fixed point in space, and that the increasing pressure should be evenly distributed, more laterally over the heels and more medially over the forefoot. In addition, the area of contact between the body and the seat of the stool should be maintained, although it diminishes a little at the back and the pressure under the right and left ischial tuberosities decreases while that on the dorsal aspects of the thighs increases towards the front edge of the seat. The *limitatio* ensures that the longitudinal arches of the feet are actively braced as the strain upon them increases. The incipient supporting function arising in the legs and the forward inclination of the long axis of the body are limited, because the contact between the body and the stool has to be maintained.

In the example starting from standing upright on the left leg, in which raising the right leg to the horizontal has already been actively buttressed at the left hip joint and in the lumbar spine by the *conditio,* a further *conditio* designates DP left greater trochanter as a fixed point in space. The *limitatio* prevents the left greater trochanter from being displaced to the left, and this constitutes activated passive buttressing.

Conditio Regulating the Speed of Movement – *Limitatio* of Economical Activity by Finding the Optimum Speed

● *Examples*
For the exercise of inclining the long axis of the body forwards by flexion at the hip joints while sitting on a stool, a speed of '30 forward bends per minute' is the optimum or ideal for the to-and-fro movement. The movement sequence is carried out at a brisk, steady pace without the need for braking activity to maintain the contact between the body and the stool.
The optimum speed in raising one leg extended while standing upright on the other is 2 seconds for raising the leg, 1 second for lowering it and 1 second for resting. This is a rate at which it is possible to observe the many demands set by the *conditio.*

4.2 Formula for Instructions in Patient Language

4.2.1 Instructions Appealing to the Patient's Perceptions

The physiotherapist indicates by words, gestures or manipulation:
- Two reference points on the patient's body, the distance between which should be increased or decreased; both points may be moved, or just one. In this way, movement excursions can be selectively initiated at particular joints. This kind of movement instruction is not specific as to the spatial orientation of the movement or the muscular activities which it evokes.
- Topographically circumscribed skin zones which the patient is asked to 'iron out' or crease. Such movement instructions call for the activation of certain muscles which contract when creasing the skin and extend when smoothing it out.
- Fixed or moving points in the patient's environment which distance points on the patient's own body should move towards or away from, or which are to be touched, or on which the patient should exert or reduce pressure.
- Directions of movement for distance points on the patient's own body, orientated according to the force of gravity, i. e. up and down. Such movement instructions ini-

tiate lifting and braking-lowering of the weight of the body, body segments or parts of body segments.
- Directions of movement for distance points on the patient's own body which utilize the patient's orientation outwards from his own body and move the whole body or parts of it forwards, backwards, left and right, and precisely determine the relation of the body or parts of it to the support area, including whether the support area remains constant or changes.
- Points on the patient's body which can be guided in a specific direction defined in terms of the environment, the support area or orientation outwards from the body, while other points on the body remain still (fixed points) or move in a different direction, thus causing alterations in the distance between points on the body. Such movement instructions can initiate very precisely differentiated, exactly planned movement sequences.
- Images which appeal to the patient's imagination and natural instincts for playing and acting.
- Tunes and rhythms, which influence the speed of movement and thus the intensity of economical activity and the degree of muscular stress.

4.2.2 Verbal – Perceptual Instruction

For verbal instructions, it is necessary that the patient is capable of comprehending language and understands the physiotherapist's idiom. A properly formulated instruction can be carried out without any difficulty, as long as it is within the patient's physical ability. The physiotherapist should therefore constantly critically review the way she formulates her verbal instructions.

- *Example*
'Put your right index finger into your right ear and your left index finger into your left ear. Use your thumbs to touch the outside of the cartilagenous part of the ears. Pinch the ear cartilage between your thumbs and index fingers. Now flap the ears out so that they are at right angles to your head, like 'jug ears'. Then pull them slowly and steadily sideways, backwards and upwards, away from the back of your throat. It almost hurts. Then let go slowly and wait. After about 30 seconds you will feel a gentle but quite definite warmth in the inner ear. When that happens, the exercise has achieved its desired effect. You should do this exercise twice an hour and whenever you start to feel dizzy.'

4.2.3 Manipulative – Perceptual Instruction

For instruction by manipulation, it is necessary that the physiotherapist's hands make contact with the patient to facilitate a movement sequence but without changing it in any essential. When the aim is to improve, by physical contact, the patient's perception of one or several distance points or their relationship to each other or the base support, manipulation can hardly fail. However, if the therapist assumes the weight of parts of the patient's body in order to facilitate perception of unfamiliar postures or movements, it becomes more complex: the counterweight to the part of the weight which the therapist has taken over has to be reduced correspondingly, so that the patient's perception of the distribution of the equilibrium is not interfered with. For this reason, manipulative limitation of an unwanted continuing movement must be neutralized by a countermanipulation. The process of learning a movement sequence can be considerably eased and speeded up by this kind of manipulation.

- *Example*
We refer to the first phase of the Dreaming Traffic Policeman exercise, with the patient lying on his left side (see *Therapeutic Exercises*). The left leg and the right arm are accelerating weights, being in front of the bisecting plane, which is identical with the midfrontal plane of BS thorax and BS head. The right leg and part of the pelvis are braking weights. The weights of the thorax, head and left arm are neutral; they are in line with the bisecting plane. The backwards tilt of the left arm and the head stabilizes the long axis of the body in slight extension. The head and both arms and legs are in free play. If the patient is unable to maintain equilibrium in this position, which involves high-intensity economical activity, it is advisable to reduce the lifting strain imposed on the patient. This can best be done by providing a supportive device for all five extremities. If the physiotherapist is at least as tall as the patient, her own body can provide the support needed, otherwise any available wall space is suitable. The lift demanded of the muscles is thus reduced while the intensity of the buttressing muscle activity is increased.

4.3 Adapting an Exericse to the Patient's Constitution and Condition

When the therapist has selected an exercise, and a functional status report is available (see p. 213), crude adaptation is carried out automatically, since the therapist can see from the report what she can ask of the patient in his present state and what constitutional factors will have to be accepted as unalterable.

> **Note**
> Nevertheless, during practice of an exercise, unexpected problems can arise which are best overcome by *not* being confronted at first. Practising an exercise is helpful only if you stay within the bounds of the possible.

4.3.1 Role of Lengths, Widths, Depths and Distribution of Weights in Adapting an Exercise

Irrespective of whether the direction components of the critical distance point of the primary movement are predominantly vertical or horizontal, the physiotherapist must bear in mind the unalterable givens of the patient's constitution. The weights to be raised and lowered must be assessed and the probable equilibrium reactions foreseen. The therapist should never judge a patient by her own condition or make comparisons between one patient and another.

- *Examples*
Bending down (see p. 281). For bending down, we choose as the critical distance point of the primary movement a point of the body which has purely vertical direction components during the movements of bending down and straightening up again. However, we must also take into account the weights of the inherently mobile system of the body which move out horizontally forwards and backwards and then return, ensuring economical bending behaviour.

Reactive arm swing (see p. 282). In this movement, the critical distance point has a purely horizontal direction component.

4.3.2 Common Causes of Error in Adapting an Exercise to the Patient's Condition

- Lack of fitness or motivation to improve fitness can often be influenced by adapting excercises to suit the patient's condition.
- Pain occurring during an exercise, particularly if it persists immediately after a certain movement, is a contraindication for that exercise.
- Weakness and inertia in muscular reaction demand patience and, in adaptation, a reduction in the strain placed upon the muscles and – if need be – reduction in the scope of the movement. The speed of movement must be chosen with great care.
- A great deal of control is needed when dealing with either restricted mobility or hypermobility. If hypermobility is present, a special effort should be made to select exercises in which the joints can be stabilized in position before reaching the end of their movement tolerance. The technique of active buttressing is often successful in cases of restricted mobility.
- Disorders of the central nervous system require appropriate treatment techniques. Basically, one has to select exercises which do not demand too high precision.

5 Functional Measuring

The physiotherapist's eyes, ears and hands are more important than protractor and tape measure when measuring a patient. Even she has a special aptitude for this task, she must constantly practise measuring by eye and keep her hearing and sense of touch sensitive and finely tuned.

Of course, protractors and tape measures are useful aids in the hands of the physiotherapist, particularly when measuring is always carried out by the same person. We all know how measurements can vary and how they can then be wrongly interpreted to mean that the patient has progressed or regressed.

> **Note**
> Reference is here made to H. U. Debrunner's measuring technique, the Neutral-0-Method, which is used in functional kinetics and takes the upright human posture as the criterion of the zero position.

● *Example*
50° Flexion, 10° extension is expressed as 'FLEX/EXT 50-0-10'.
50° Flexion to 7° flexion contracture is expressed as 'FLEX/EXT 50-7-0'.

> **Note**
> As the designations FLEX/EXT/ABD/ADD/IR/ER/ + ROT/ − ROT are used to denote both static positions of joints and dynamic movements at joints, we use the words 'flexional/extensional/abductional/adductional/rotational/translational' when speaking of movement. This allows us to define a movement in, say, the field of flexion of the hip joint precisely according to its direction either as flexional or extensional. The *position* means the angle formed by two levers or pointers within the field of a movement component, the size of which is measured in relation to their zero position.

System of Functional Measurement

Body segment by body segment, we set up reference *points, lines* and *axes,* which we use in functional analysis, in the functional status report, and in instructing the patient.
The *points* are the distance points which, moving towards each other or away from each other, indicate the quality of the movement.

157

The *lines* lengthen or shorten, smoothing or creasing the skin, and thus indicating particular movements at the joints.

The *axes,* if they are levers or rotating pointers, show the precise position of the joints. If they are straight lines of orientation, like the long axis of the body, the diameters of the thorax, the lengths of body segments or of their parts, or axes of motion, they show by their positional relationship whether the joints are at zero or how far from it they have moved.

The value of this system of measurement lies in the fact that each measurement can be verified by three different sets of data and that, during observation of movement, alterations of joint positions within the body can be registered according to various different signs.

We begin with *BS thorax,* which has been defined as the stabile (see p. 80). Then follows *BS head,* the unpaired extremity, positioned in the plane of symmetry. *BS pelvis* and *BS legs* we take together, because the hip joints which link them must be regarded as part of both. That these segments belong together functionally is due to the dominant role of the hip joints in locomotion. The last segment is *BS arms.*

5.1 Points, Lines and Axes of BS Thorax

Normal appearance in upright stance in the zero position: The intrinsically mobile thoracic spine, when dynamically stabilized at zero, is slightly curved in the plane of symmetry (Fig. 187). The convex curve faces backwards (normal thoracic kyphosis). The ribs incline from behind and above towards the front and downwards. The obliquity increases caudally. A characteristic of the even curvature of the ribs is that the frontotransverse diameter of the thorax is greater than the sagittotransverse diameter. The sternum is inclined downwards and a little backwards.

Flat back: A variant within the norm. The normal kyphosis of the thoracic spine is reduced (Fig. 188). The frontotransverse diameter of the thorax is greater, the sagittotransverse smaller than in the norm. The curvature of the ribs is less even. It is difficult to stabilize arm movements with adduction components at the shoulder joints, because the contact between the shoulder blades and the wall of the thorax is incomplete. Total flattening of the thoracic curve and/or a funnel chest must be considered pathological.

Round back: A variant within the norm. The normal kyphosis of the thoracic spine is increased (Fig. 189). The frontotransverse diameter of the thorax is smaller, the sagittotransverse diameter greater than in the norm. The curvature of the ribs is less even. It is difficult to stabilize arm movements with abduction components at the shoulder joints, because the contact between the shoulder blades and the wall of the thorax is incomplete. The thoracic kyphosis gives rise to compensatory increased lordosis of the lumbar and cervical spines, resulting in hyperkypholordosis. Markedly increased thoracic kyphosis and/or a pigeon breast are pathological.

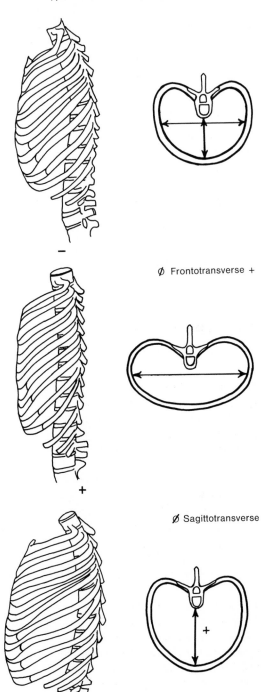

Fig. 187. BS thorax (norm)

Ø Frontotransverse +

Fig. 188. BS thorax (flat-back variant
of the norm)

Ø Sagittotransverse

Fig. 189. BS thorax (round-back variant of the norm)

The joints of BS thorax, the movements of which are functionally measured by the changes in position and shape of points, lines and axes of the body, are (Fig. 190):

– The 12 motion segments of the thoracic spine (12 pairs of vertebral joints and 12 intervertebral discs)
– 12 pairs of ribs with their costovertebral joints and costosternal junctions

BS thorax articulates with BS arms at the sternoclavicular joints, BS head at motion segment C7/T1 and BS pelvis at motion segment T12/L1.

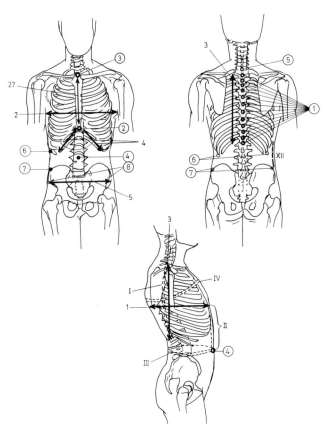

Fig. 190. Points, lines and axes on BS thorax. *Points:* ① spinous processes of T1–T12; ② xiphoid process; ③ suprasternal notch; ④ navel; ⑤ spinous process of C7; ⑥ right/left lower border of thorax (midfrontal plane); ⑦ right/left iliac crest (midfrontal plane); ⑧ right/left anterior iliac spine. *Lines:* I line connecting spinous processes of C7–T12; II line connecting xiphoid process and navel; III circumference of the waist at the level of the navel; IV circumference of the thorax at the level of the axilla; XII lateral lines of the waist. *Axes:* 1 sagittotransverse thoracic diameter (at the level of T6/T7); 2 frontotransverse thoracic diameter (at the level of T6/T7); 3 extent of contribution of thoracic spine to the total body length (distance T1–T12); 4 epigastric angle; 5 line connecting the anterior iliac spines; 27 long axis of sternum

160

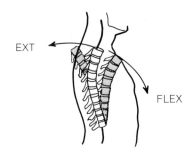

EXT

FLEX

Fig. 191. Flexion/extension of the thoracic spine

5.1.1 Movements of the Thoracic Spine in the Plane of Symmetry (Fig. 191)

Flexion: The curvature in the plane of symmetry increases.

Points: DP xiphoid process ② and DP navel ④ move towards each other. Usually ② moves towards ④ and the upper abdomen 'becomes shorter'.

Lines: The line connecting the spinous processes of C7–T12 (I) becomes longer. The line connecting the xiphoid process and the navel (II) becomes shorter.

Axes: The proportion of the total body length contributed by the thoracic spine (3, the long axis of the thorax) decreases; the frontotransverse diameter of the thorax at the level of T5/T6 (2) and the line connecting the anterior iliac spines (5) remain parallel to each other.

Extension: The curvature in the plane of symmetry decreases.

Points: DP xiphoid process ② and DP navel ④ move away from each other. Usually ② moves away from ④ and the upper abdomen 'becomes longer'.

Lines: The line connecting the spinous processes of C7–T12 (I) becomes shorter; the line connecting the xiphoid process and the navel (II) becomes longer.

Axes: The proportion of the total body length contributed by the thoracic spine (3) increases; the frontotransverse diameter of the thorax at the level of T5/T6 (2) and the line connecting the anterior iliac spines (5) remain parallel to each other.

5.1.2 Movements of the Thoracic Spine in the Frontal Plane (Fig. 192)

Lateral flexion: Right/left concave curvature in the frontal plane

Points: DP right and DP left border of thorax in the midfrontal plane ⑥ and DP right and DP left border of the pelvis in the midfrontal plane ⑦ move further apart on the convex side and closer together on the concave side; ⑥ moves, ⑦ is stationary.

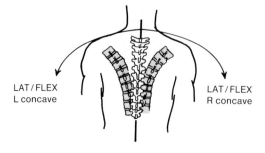

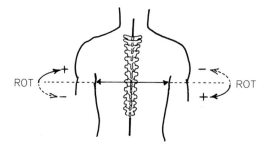

Fig. 192. Right/left concave lateral flexion of the thoracic spine

Fig. 193. Positive/negative rotation of the frontotransverse thoracic diameter at the level of rotation of the thoracic spine

Lines: The connecting line C7–T12 forms a right/left concave arch and lengthens.

Axes: The frontotransverse diameter of the thorax (2) and the line connecting the anterior iliac spines (5) form an angle at the point where they meet on the concave side of the curvature. 2 moves in the midfrontal plane while 5 remains frontotransverse.

5.1.3 Movements of the Thoracic Spine in the Transverse Plane (Fig. 193)

Rotation in the thoracic spine: +(positive)/−(negative) rotation of the frontotransverse diameter of the thorax (cranial rotary pointer) about the long axis of the body in a transverse plane; i.e. positive/negative rotation of the thorax in the thoracic spine.

Points: DP xiphoid process ② performs a positive rotation in a transverse plane. Its distance from the left iliac spine increases, its distance from the right iliac spine decreases. If it performs a negative rotation in a transverse plane, the distances change in the reverse way.

Lines: The line connecting the xiphoid process and the navel (II) becomes longer.

Axes: The proportion of total body length contributed by the thoracic spine remains unaltered. The rotary pointer, the frontotransverse diameter of the thorax (2), rotates about the long axis of the body and forms an angle with the line connecting the iliac spines (5). If 2 is projected onto 5, the angle of rotation can be measured. If 2 has performed a positive rotation, +ROT of the thorax in the thoracic spine has taken place; −ROT occurs when a negative rotation is performed.

5.1.4 Movements of the Thorax in Transverse Planes (Figs. 194, 195)

Translations of the thorax in the lower thoracic spine/upper lumbar spine laterally to the right/left and ventrally/dorsally.

Points: DP suprasternal notch ③ and DP xiphoid process ② together move ventrally or dorsally, or laterally to the left or right, away from DP navel. The whole of the thorax is translated en bloc.

Lines: The right and left lateral lines of the waist (XII) become longer as DPs right and left lower border of thorax in the midfrontal plane ⑥, in relation to DPs right and left iliac crest in the midfrontal plane ⑦, move right during a translation to the right, left during a translation to the left, ventrally in a ventral translation and dorsally in a dorsal translation.

Axes: While the frontotransverse diameter of the thorax (2) moves right or left, ventrally or dorsally in relation to the line connecting the anterior iliac spines (5), the two axes

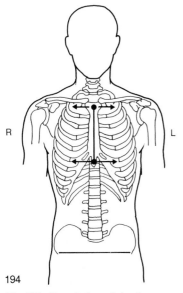

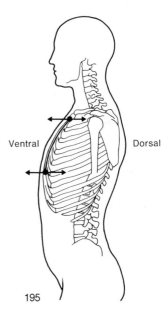

R   L   Ventral   Dorsal

194   195

Fig. 194. Translation of the thorax to the right/left

Fig. 195. Ventral/dorsal translation of the thorax

remain parallel. The long axis of the sternum (27) glides along a parallel path during translation, flexing neither to right nor left in lateral translation nor forwards/backwards in ventral/dorsal translation.

5.1.5 Movements of the Ribs in Inspiration and Expiration (Figs. 196, 197)

Lifting the ribs: Widening of the intercostal spaces; inspiration.

Points: DP suprasternal notch ③ moves ventrally/cranially. DP navel ④ moves ventrally. DPs right and left lower border of thorax in the midfrontal plane ⑥ move laterally and cranially.

Lines: The waistline (III) expands, as does the chest circumference at the level of the axilla (IV).

Axes: The sagittotransverse (1) and the frontotransverse (2) diameters of the thorax lengthen and the epigastric angle (4) increases. The proportion of the total body length contributed by the thoracic spine (3) remains unchanged.

Lowering the ribs: Narrowing of the intercostal spaces; expiration.

Points: DP jugular notch ③ moves dorsally/caudally. DP navel ④ moves dorsally and DPs right and left borders of the thorax in the midfrontal plane ⑥ move medially/caudally.

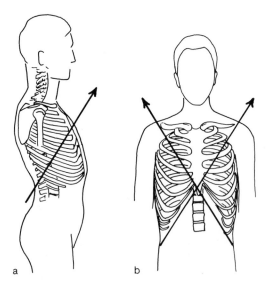

Fig. 196a, b. Inspiratory lifting of the ribs: **a** lateral aspect, **b** ventral aspect

a   b

164

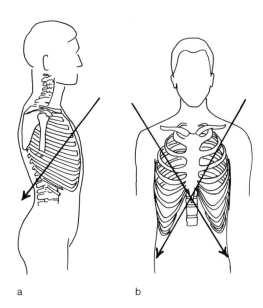

Fig. 197 a, b. Expiratory lowering of the ribs: **a** lateral aspect, **b** ventral aspect

a   b

Lines: The waistline (III) contracts, as does the chest circumference at the level of the axilla (IV).

Axes: The sagittotransverse (1) and the frontotransverse (2) diameters of the thorax decrease. The epigastric angle (4) is reduced. The proportion of the total body length contributed by the thoracic spine (3) remains unchanged.

5.2 Points, Lines and Axes of BS Head

Normal appearance in upright stance in the zero position: BS head is aligned in the long axis of the body, which is vertical. The eyes are horizontal and face forwards. The cervical spine and atlanto-occipital and atlanto-axial joints are at zero and potentially mobile. The cervical spine is slightly curved in the plane of symmetry, the convex aspect facing forwards (cervical lordosis; Fig. 198).

Flat back: A variant within the norm. DP vertex is aligned in the long axis of the body. The lordotic curvature of the cervical spine in the plane of symmetry is reduced. The cervical spine and the atlanto-occipital and atlanto-axial joints are potentially mobile. To allow the eyes to be horizontal and face forwards, there is slightly increased extension at these joints (Fig. 199). Any forward deviation of the head out of the long axis of the body should be considered pathological, as this reduces the potential mobility of the cervical spine. Total flattening out of the physiological cervical lordosis and any ky-

165

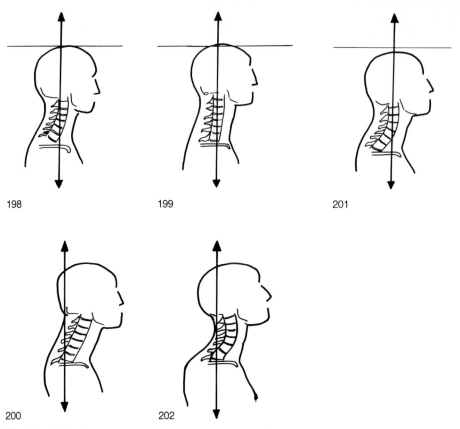

Fig. 198. BS head (norm)

Fig. 199. BS head (flat-back variant of the norm)

Fig. 200. Total flattening out of the cervical lordosis

Fig. 201. BS head (round-back variant of the norm)

Fig. 202. Hyperlordosis of the cervical spine

phosis in the lower cervical spine (cervical or neck kyphosis; see p. 252) must also be considered pathological (Fig. 200).

Round back: A variant within the norm. DP vertex is aligned in the long axis of the body. The lordotic curvature of the cervical spine in the plane of symmetry is more pronounced. The cervical spine and the atlanto-occipital and atlanto-axial joints are potentially mobile. To allow the eyes to be horizontal and face forwards, there is slightly increased flexion at these joints (Fig. 201).

Any forward deviation of the head out of the long axis of the body and a marked increase in the lordotic curvature of the neck, particularly in the middle region, must be considered pathological, as must slipping of C5 on C6 (Fig. 202).

166

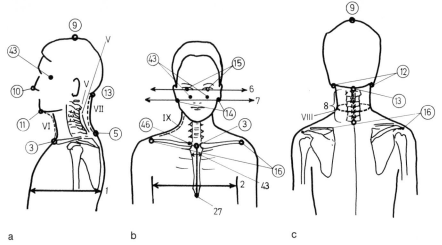

a   b   c

Fig. 203a–c. Points, lines and axes on BS head. *Points:* ③ suprasternal notch; ⑤ spinous process of C7; ⑨ vertex; ⑩ tip of the nose; ⑪ point of chin; ⑫ right/left mastoid process; ⑬ external occipital protuberance; ⑭ right/left earlobe; ⑮ right/left eye; ⑯ right/left acromioclavicular joint; ㊸ right/left zygomatic arch; ㊻ right/left sternal end of clavicle; ㊼ articular mandibular process (see Fig. 209). *Lines:* V line connecting the spinous processes of C1–C7; VI ventral line of the neck from ⑪ to ③; VII dorsal line of the neck from ⑬ to ⑤; VIII circumference of the neck at the level of the larynx; IX right/left lateral line of the neck from ⑭ to ⑯; *Axes:* 6 line connecting the eyes; 7 line connecting the earlobes; 8 extent of contribution of the cervical spine to the total body length (distance C1–C7); 27 long axis of sternum; 43 transverse axis of manubrium sterni

The joints of BS head, the movements of which are functionally measured by measuring the changes in position and shape of points, lines and axes of the body, are (Fig. 203):

The 7 motion segments of the cervical spine (7 pairs of vertebral joints and the atlanto-occipital and atlanto-axial joints, and 6 intervertebral discs) and the temporomandibular joints.

BS head articulates with BS thorax at the level of motion segment C7/T1.

5.2.1 Movements of the Cervical Spine in the Plane of Symmetry (Fig. 204)

Flexion: The curvature in the plane of symmetry decreases and the head bends forwards.

Points: DP point of chin ⑪ and DP suprasternal notch ③ move closer together. DP external occipital protuberance ⑬ and DP spinous process of C7 ⑤ move further apart. Usually, ⑪ moves closer to ③, which remains stationary.

167

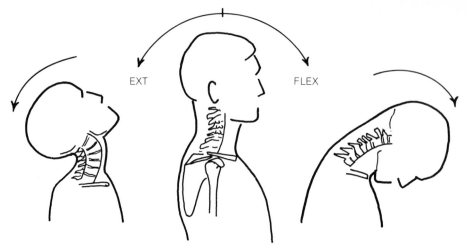

Fig. 204. Flexion/extension of the cervical spine

Lines: The ventral line of the neck from the point of the chin to the suprasternal notch (VI) becomes shorter. The dorsal line of the neck from the external occipital protuberance to the spinous process of C7 (VII) becomes longer. The line connecting the spinous processes of C1–C7 (V) becomes longer.

Axes: The line connecting the eyes (6) and the frontotransverse diameter of the thorax (2) remain parallel.

Extension: The curvature in the plane of symmetry increases and the head bends backwards.

Points: DP point of chin ⑪ and DP suprasternal notch ③ move further apart. DP external occipital protuberance ⑬ and DP spinous process of C7 ⑤ move closer together. Usually, ⑪ moves away from ③, which remains stationary.

Lines: The ventral line of the neck from the point of the chin to the suprasternal notch (VI) becomes longer. The dorsal line of the neck from the external occipital protuberance to the spinous process of C7 (VII) becomes shorter, as does the line connecting the spinous processs of C1–C7 (V).

Axes: The line connecting the eyes (6) and the frontotransverse diameter of the thorax (2) remain parallel.

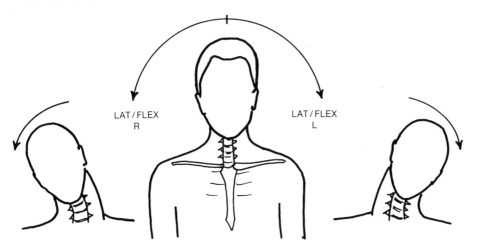

Fig. 205. Right/left concave lateral flexion of the cervical spine

5.2.2 Movements of the Cervical Spine in the Frontal Plane (Fig. 205)

Lateral flexion: Right/left concave curvature in the frontal plane; the head inclines sideways.

Points: DP right mastoid process ⑫ and DP right acromioclavicular joint ⑯ move closer together, while DP left mastoid process ⑫ and DP left acromioclavicular joint ⑯ move further apart and DP vertex ⑨ moves laterally/right/caudally (a right concave movement in the cervical spine). The opposite happens for lateral concave curvature to the left.

Lines: The right/left lateral line of the neck (IX) between the right/left ear lobe and the right/left acromioclavicular joint becomes shorter on the concave and longer on the convex side.

Axes: The line of the frontotransverse diameter of the thorax (2) and the line connecting the eyes (6) form an angle where they meet on the concave side. The line connecting the eyes (6) and that connecting the ear lobes (7) move in a frontal plane.

5.2.3 Movements of the Cervical Spine and the Atlanto-occipital and Atlanto-axial Joints in Transverse Planes (Fig. 206)

Rotation in the cervical spine and at the atlanto-occipital and antlanto-axial joints: +(positive)/ −(negative) rotation of the line connecting the ear lobes (cranial rotary pointer) about the long axis of the body in a transverse plane.

169

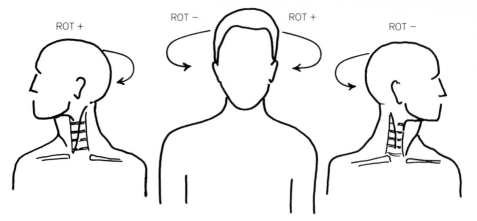

Fig. 206. Positive/negative rotation of the head in the cervical spine

Points: DP point of chin ⑪ and DP right acromioclavicular joint ⑯ move further apart; DP point of chin ⑪ and DP left acromioclavicular joint ⑯ move closer together (=ROT CS/atlanto-occipital and atlanto-axial joints; the cranial pointer performs a negative rotation). The same rotation can be performed by the caudal pointer rotating positively. The position of DP vertex ⑨ remains unchanged.

Lines: The ventral line of the neck from the point of the chin to the suprasternal notch (VI) and the dorsal line of the neck from the external occipital protuberance to the spinous process of C7 (VII) both become longer.

Axes: The proportion of the total body length contributed by the cervical spine (i. e. 8, the length C1–C7), remains about the same. The line connecting the ear lobes (7) rotates about the long axis of the body in a transverse plane. The line connecting the ear lobes (7) forms an angle with the frontotransverse diameter of the thorax (2). The angle of rotation can be measured when 7 is projected onto 2.

5.2.4 Movements of the Head in Transverse Planes (Figs. 207, 208)

Ventral translation of the head at the level of the cervical spine.

Points: DP external occipital protuberance ⑬ and DP spinous process of C7 ⑤ move further apart, ⑬ moving ventrally in the plane of symmetry; or DP point of chin ⑪ and DP suprasternal notch ③ move further apart, ⑪ moving ventrally in the plane of symmetry. DP vertex ⑨ moves ventrally in the plane of symmetry. The head is thus translated ventrally as a gliding body, the eyes still facing forward.

170

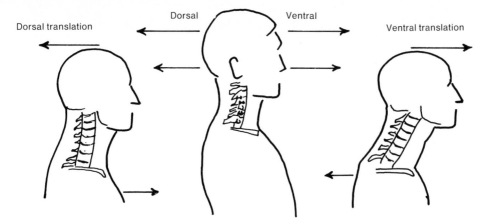

Fig. 207. Ventral/dorsal translation of the head

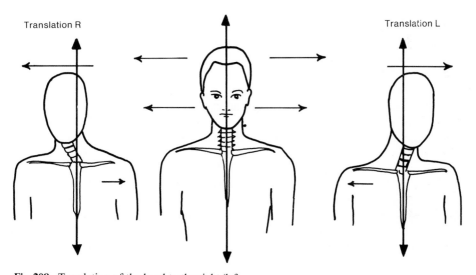

Fig. 208. Translation of the head to the right/left

Lines: The ventral line of the neck from the point of the chin to the suprasternal notch (VI) and the dorsal line of the neck from the external occipital protuberance to the spinous process of C7 (VII) become longer.

Axes: The connecting lines between the ear lobes (7) and between the eyes (6) remain parallel to the frontotransverse diameter of the thorax (2). 7 and 6 move ventrally while 2 remains fixed.

171

Dorsal translation of the head at the level of the cervical spine.

Points: DP external occipital protuberance ⑬ moves dorsally, closer to DP spinous process of C7 ⑤ in the plane of symmetry. DP vertex ⑨ and DP point of chin ⑪ also move dorsally, closer to DP suprasternal notch ③. The head is thus translated dorsally as a gliding body, the eyes still facing forward.

Lines: The ventral line of the neck from the point of the chin to the suprasternal notch (VI) becomes shorter, as does the dorsal line of the neck from the external occipital protuberance to C7 (VII).

Axes: The connecting lines between the ear lobes (7) and between the eyes (6) remain parallel to the frontotransverse diameter of the thorax (2). 7 and 6 move dorsally while 2 remains fixed.

Right/left lateral translation of the head at the level of the cervical spine.

Points: DP right mastoid process ⑫ moves laterally to the right, towards DP right acromioclavicular joint ⑯. DP right mastoid process ⑫ and DP right acromioclavicular joint ⑯ thus move closer together, while DP left mastoid process ⑫ and DP left acromioclavicular joint ⑯ move further apart (or, for left lateral translation, the other way around). The head thus undergoes right/left translation as a gliding body, the eyes still facing forward.

Lines: In right lateral translation of the head, the left lateral line of the neck from the left ear lobe to the left acromioclavicular joint (IX) becomes longer and the right shorter; for left lateral translation, the opposite occurs.

Axes: In a translation of the head to the right the connecting lines between the eyes (6) and between the ears (7) remain parallel to the frontotransverse diameter of the thorax (2). 6 and 7 glide in a transverse plane to the right of the sagittotransverse diameter of the thorax (1); for left lateral translation, the opposite occurs.

Note
Translations of the head in the upright posture are finely differentiated, straight, economical movements in the eye–ear plane, the so-called Frankfurt horizontal.

- *Examples*

Translation of the head	Forwards	Backwards
Critical distance point	Right/left zygomatic arch ㊸	Right/left zygomatic arch ㊸
Direction of movement	Forwards	Backwards
Spatial path	Straight line	Straight line
Critical pivot	Motion segment C6/C7	Motion segment C6/C7

172

Movement excursion	Co-rotational continuing movement	Co-rotational continuing movement
At the atlanto-axial and atlanto-occipital joints	Extensional	Flexional
In the cervical spine	Translational	Translational
Buttressing	Active buttressing, thoracic spine at zero/extensional	Active buttressing, thoracic spine at zero/flexional
Translation of the head	*To the right*	*To the left*
Critical distance point	Right/left ear lobe ⑭	Right/left ear lobe ⑭
Direction of movement	To the right	To the left
Spatial path	Straight line	Straight line
Critical pivot	Motion segment C6/C7	Motion segment C6/C7
Movement excursion	Co-rotational continuing movement	Co-rotational continuing movement
At the atlanto-axial and atlanto-occipital joints	Lateroflexional/left concave	Lateroflexional/right concave
In the cervical spine	Translational	Translational
Buttressing	Active buttressing, thoracic spine at zero/lateroflexional left concave	Active buttressing, thoracic spine at zero/lateroflexional right concave

5.2.5 Movements at the Temporomandibular Joints (Figs. 209, 210)

Biting: Opening and closing the mouth (hinge joint).

Points: Opening the mouth: DP point of nose ⑩ and DP point of chin ⑪ move further apart, DP point of chin ⑪ moving caudally/dorsally in the plane of symmetry. Closing the mouth: DP point of nose ⑩ and DP point of chin ⑪ move closer together, DP point of chin ⑪ moving cranially/ventrally in the plane of symmetry.

Lines: In opening the mouth the ventral line of the neck between the point of the chin and the suprasternal notch (VI) becomes longer; in closing the mouth it becomes shorter.

Axes: In opening the mouth the angle formed between the lines connecting the point of the chin with the right and left articular mandibular process and the ear–eye plane increases.

Grinding: Movement of the mandible to the right and the left (rotary-translatory joint).

Points: DP point of chin ⑪ and DP point of nose ⑩ move further apart. In grinding to the right, DP point of chin ⑪ moves towards DP right zygomatic arch ㊸ and away from DP left zygomatic arch ㊸; in grinding to the left, the opposite occurs.

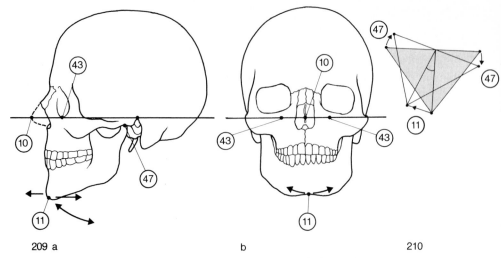

209 a   b   210

Fig. 209 a, b. Movements of the mandibular joints: biting, grinding and gnawing

Fig. 210. Isosceles triangle formed by the point of the chin and the right/left articular mandibular processes. Craniocaudal view

Lines: In grinding movements to the right, the line connecting DP right ear lobe ⑭ and DP point of chin ⑪ becomes shorter while the line connecting DP left ear lobe ⑭ and DP point of chin ⑪ becomes longer; the reverse happens in grinding movements to the left.

Axes: The altitude of the isosceles triangle formed by DPs point of chin ⑪ and right and left articular mandibular processes 47 forms an angle with the plane of symmetry in grinding movements to either left or right; the angle is closed dorsally.

Gnawing: Ventral and dorsal gliding of the mandible (translatory joint).

Points: Ventral gliding of the mandible: DP point of chin ⑪ and DPs right and left mastoid processes ⑫ move further apart.
Dorsal gliding of the mandible: DP point of chin ⑪ and DPs right and left mastoid processes ⑫ move closer together, DP point of chin ⑪ moving dorsally in the plane of symmetry.

Lines: In dorsal gliding of the mandible the two lines connecting DP point of chin ⑪ with DPs right and left ear lobe ⑭ become longer; in ventral gliding they become shorter.

Axes: The mandibular teeth glide in a movement that keeps them parallel to the maxillary teeth. In ventral gliding of the mandible the mandibular teeth move ventral of the maxillary teeth, in dorsal gliding they move dorsal of the maxillary teeth.

174

5.3 Points, Lines and Axes of BS Pelvis and BS Legs

Normal appearance in the upright stance in the zero position: The long axes of BS pelvis, BS thorax and BS head form the long axis of the body. In upright stance in the neutral position the long axis of the body is vertical and BSs pelvis, thorax and head are

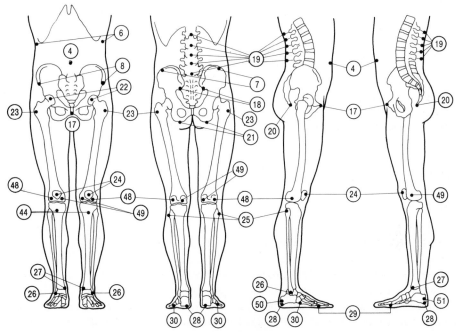

Fig. 211. Points on BS pelvis and BS legs: ④ navel; ⑥ right/left lower border of thorax (midfrontal plane); ⑦ iliac crest (midfrontal plane); ⑧ right/left anterior iliac spine; ⑰ symphysis pubis; ⑱ right/left dimple over the sacro-iliac joints; ⑲ spinous processes of L1–L5; ⑳ point of coccyx; ㉑ right/left ischial tuberosity; ㉒ right/left pulsation point of the femoral artery (to aid localization of the right/left femoral head); ㉓ right/left greater trochanter of the femur; ㉔ right/left patella (middle or cranial pole); ㉕ right/left fibular head; ㉖ right/left lateral malleolus; ㉗ right/left medial malleolus; ㉘ right/left point of the calcaneus; ㉙ right/left metatarsophalangeal joints of the toes I–V; ㉚ right/left metatarsal head V; ㊹ right/left tibial tuberosity; ㊽ right/left lateral femoral epicondyle; ㊾ right/left medial femoral epicondyle; ㊿ right/left tubercle of lateral calcaneal tuberosity; ㊿① right/left tubercle of medial calcaneal tuberosity

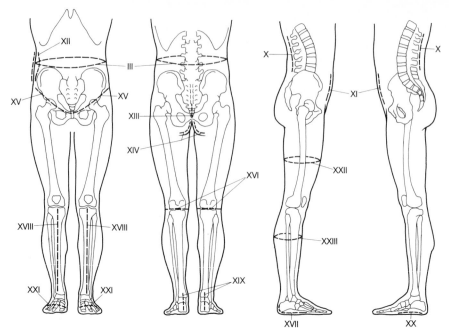

Fig. 212. Lines on BS pelvis and BS legs: III circumference of the waist at the level of the navel; X line connecting spinous processes of L1–L5; XI line connecting the navel and symphysis pubis; XII right/left lateral line of the waist; XIII anal fold; XIV right/left gluteal fold; XV right/left skin fold of the groin; XVI right/left popliteal fold of the knee joint; XVII right/left lateral border of the foot; XVIII right/left anterior border of the tibia; XIX right/left Achilles' tendon (to assess the position of the heel); XX right/left longitudinal arch of the foot; XXI right/left transverse arch of the foot; XXII circumference of the right/left thigh (the level must be stated); XXIII circumference of right/left lower leg (mid-calf)

aligned in it. BS thorax is the stabile if the thoracic spine is dynamically stabilized at zero. BS head has potential mobility in the cervical spine and at the atlanto-occipital and atlanto-axial joints. The pelvis is in the two-legged stance position and potentially mobile, particularly about the flexion/extension axes of the joints of hip and lumbar spine. The convex face of the lumbar lordosis faces forwards. The pelvis is balanced over the femoral heads, each thigh over the tibial plateau, each lower leg over the talus and the latter over the calcaneum, which forms the dorsal pillar of the longitudinal arch of the foot, whose functional long axis points forward (see p. 265). Neither hips nor knees are at end-stop. BS legs also has potential mobility and is in fact a mobile.

The joints of BS pelvis and BS legs, the movement excursions of which can be functionally measured by measuring the changes in position and shape of points, lines and axes, are as follows (Figs. 211–213):

The five motion segments of the lumbar spine (five pairs of vertebral joints and five intervertebral discs, the most caudal of which forms the lumbosacral articulation;

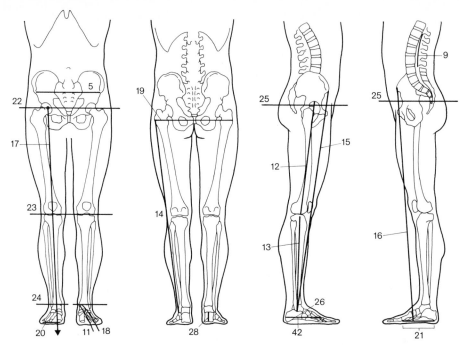

Fig. 213. Axes on BS pelvis and BS legs: 9 Extent of contribution of lumbar spine to the total body length (distance L1–L5); 11 right/left anatomical long axis of the foot (from the centre of the heel to the second toe ray); 12 long axis of the right/left thigh; 13 right/left long axis of lower leg; 14 right/left anatomical length of leg (distance from greater trochanter to lateral malleolus); 15 right/left clinical length of leg (distance from anterior iliac spine to lateral or medial malleolus); 16 right/left functional length of leg (distance from anterior iliac spine to ground); 17 right/left weight-bearing axis of leg; 18 right/left axis of pronation/supination (from the centre of the heel to the third toe ray); 19 line connecting the right and left greater trochanters; 20 right/left base of transverse arch of foot; 21 right/left base of longitudinal arch of foot; 22 right/left flexion/extension axis of hip joint; 23 right/left flexion/extension axis of knee joint; 24 right/left axis of plantar flexion and dorsal extension of the foot (talocrural joint); 25 right/left abduction/adduction axis of hip joint; 26 right/left eversion/inversion axis of foot (subtalar joint); 28 right/left height of calcaneal tuberosity; 42 right/left functional long axis of foot (from the tubercle of the lateral calcaneal tuberosity to the proximal joint of the large toe)

the sacroiliac, hip and knee joints; the talocrural and subtalar joints; the tarsometatarsal, metatarsophalangeal and interphalangeal joints.

When discussing the joints of BS legs, we differentiate between the proximal and distal distance points and use the abbreviations 'pDP' and 'dDP'. In addition, whenever necessary for observation, we shall also name the pivot or fulcrum.

BS thorax articulates with BS pelvis at motion segment T12/L1; BS pelvis articulates with the BS legs at the hip joints.

5.3.1 Movements of the Lumbar Spine in the Plane of Symmetry (Fig. 214)

Flexion: The curvature in the plane of symmetry first decreases then increases; at the end of flexion, the convex aspect faces dorsally.

Points: DP symphysis pubis ⑰ and DP navel ④ move closer together. With the thoracic spine stabilized at zero, ⑰ usually moves towards ④.

Lines: The connecting line along the spinous processes of L1–L5 (X) becomes longer. The line connecting the symphysis pubis and the navel (XI) becomes shorter.

Axes: The proportion of the total body length contributed by the lumbar spine first increases and then decreases again as the convexity faces dorsally (9). The total length of the body, measured as the long axis of the body, first increases and then decreases as the dorsal convexity becomes more pronounced. The line connecting the anterior iliac spines (5) and the line of the frontotransverse diameter of thorax remain parallel.

Extension: The curvature in the plane of symmetry increases.

Points: DP symphysis pubis ⑰ and DP navel ④ move further apart. With the thoracic spine stabilized at zero, ⑰ usually moves away from ④.

Lines: The line connecting the spinous processes of L1–L5 (X) becomes shorter. The line connecting the symphysis pubis and the navel (XI) becomes longer, and small skin creases form at the lumbosacral junction.

Axes: The proportion of the total body length contributed by the lumbar spine decreases (9). The total length of the body, measured as the long axis of the body (10), decreases. The line connecting the anterior iliac spines (5) and the line of the frontotransverse diameter of thorax (2) remain parallel.

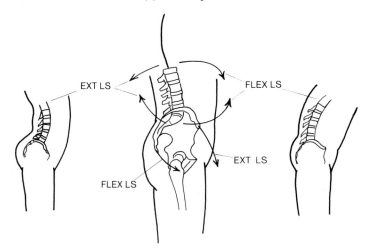

Fig. 214. Flexion/extension of the lumbar spine. *LS,* lumbar spine

5.3.2 Movements of the Lumbar Spine in the Frontal Plane (Fig. 215)

Lateral flexion: Right or left concave curvature of the lumbar spine in the frontal plane.

Points: DP lower right border of thorax in the midfrontal plane ⑥ and DP right iliac crest in the midfrontal plane ⑦ move closer together while DP lower left border of thorax in the midfrontal plane ⑥ and DP left iliac crest in the midfrontal plane ⑦ move further apart (for left lateral flexion, the opposite occurs). With the thoracic spine stabilized at zero, right DP ⑦ moves closer to the right DP ⑥, while left DP ⑦ moves away from left DP ⑥. The concavity is on the side where the DPs approach each other, the convexity on the side where they move apart.

Lines: The line connecting the spinous processes of L1–L5 (X) becomes longer.

Axes: The proportion of the total body length contributed by the lumbar spine decreases. The line connecting the anterior iliac spines (5) and the line of the frontotransverse diameter of the thorax (2) form an angle where they meet on the concave side of the curvature. 5 moves in the midfrontal plane while 2 remains frontotransverse.

5.3.3 Movements at the Hip Joints About the Frontotransverse Axis

Flexion: 180°, less the ventral angle between the long axis of the right/left thigh and the line connecting the greater trochanter ㉓ and the iliac crest (midfrontal plane) ⑦. Movement tolerance 120° (Fig. 216).

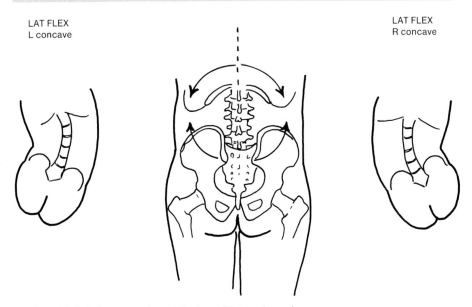

LAT FLEX
L concave

LAT FLEX
R concave

Fig. 215. Right/left concave lateral flexion of the lumbar spine

Points: pDP right/left iliac spine ⑧ and dPD right/left patella ㉔ move closer together.

Lines: The right/left fold of the groin (XV) becomes deeper while the gluteal fold (XIV) flattens out.

Axes: The line connecting the right and left iliac spines (5) and the flexion/extension axes of the right and left knee joints (23) remain parallel. The long axis of the thigh (12) and the line connecting the greater trochanter ㉓ to the iliac crest (midfrontal plane) ⑦ form an angle of less than 180°, its apex pointing dorsally.

For a weight-bearing leg, flexion of the hip joint is easiest if the fulcrum (greater trochanter) moves dorsally (backwards). With a leg in free play, flexion of the hip joint occurs from the distal lever (the thigh), very often with a co-rotational continuing movement in the form of a flexion of the lumbar spine effected by the pelvis.

Extension: 180°, less the dorsal angle between the long axis of the right/left thigh and the line connecting the greater trochanter ㉓ to the iliac crest (midfrontal plane) ⑦. Movement tolerance 15° (Fig. 217).

Points: pDP right/left iliac spine ⑧ and dDP right/left patella ㉔ move away from each other.

Lines: The right/left fold of the groin (XV) flattens out while the right/left gluteal fold (XIV) deepens.

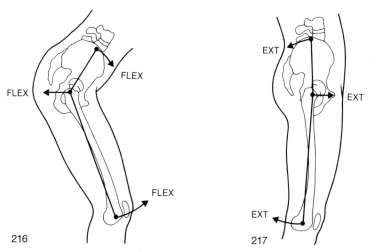

Fig. 216. Flexion at the hip joint effected by the distal and proximal distance points with displacement of the fulcrum

Fig. 217. Extension at the hip joint effected by the distal and proximal distance points with displacement of the fulcrum

Axes: The line connecting the right and left iliac spines (5) and the flexion/extension axes of the right and left knee joints ㉓ remain parallel. The long axis of the thigh (12) and the line connecting the greater trochanter ㉓ to the iliac crest (midfrontal plane) ⑦ form an angle of less than 180°, its apex pointing ventrally.

For a weight-bearing leg, extension of the hip joint is easiest if the fulcrum (greater trochanter) moves ventrally. With a leg in free play, extension of the hip joint occurs from the distal lever (the thigh), very often with a co-rotational continuing movement in the form of extension of the lumbar spine effected by the pelvis.

5.3.4 Movements at the Hip Joints About the Sagittotransverse Axis

Abduction: The medial angle in the frontal plane of the right/left hip joint formed by the long axis of the right/left thigh and the line connecting the anterior iliac spines, less 90° (zero position). Movement tolerance 50° (Fig. 218).

Points: pDP right anterior iliac spine ⑧ and dDP right patella ㉔ move toward each other laterally (the opposite happens in abduction of the left leg). If the right leg is weight bearing and the left in free play, fulcrum right hip joint moves medially, and the long axis of the right leg also inclines medially. Depending on the extent of the support area and the patient's constitution (see p. 215), the fulcrum may instead move a little laterally, with the long axis of the leg moving in the same direction. pDP right anterior iliac spine ⑧ moves caudally/laterally in relation to the right hip joint. The distal distance point of the left (free-play) leg, dDP patella ㉔, moves laterally/cranially. Fulcrum left hip joint and pdP left iliac spine ⑧ move cranially/medially in a co-rotational continuing movement (Fig. 219).

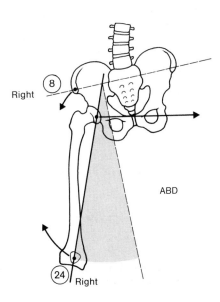

Fig. 218. Abduction at the hip joint effected by pDP ⑧ and dDP ㉔ with displacement of the fulcrum

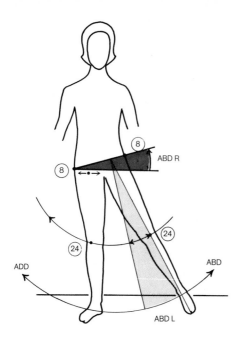

Fig. 219. Abduction at the hip joint in weight bearing

Lines: The right/left fold of the groin (XV) becomes shallower; so does the right/left gluteal fold (XIV).

Axes: The line connecting the iliac spines (5) meets the long axis of the right/left thigh (12) in the frontal plane, forming a medial angle greater than 90°.

Adduction: 90° (zero position), less the medial angle in the frontal plane of the right/left hip joint formed between the line connecting the anterior iliac spines and the long axis of the right/left thigh. Movement tolerance 30°–40° (Fig. 220).

Points: pDP left anterior iliac spine ⑧ and dDP right patella ㉔ move closer together medially (the opposite occurs in adduction of the right leg).
If the right leg is weight bearing and the left in free play, fulcrum right hip joint moves laterally, with the long axis of the right leg also moving laterally. pDP left anterior iliac spine ⑧ moves caudally/medially in relation to the right hip joint. dDP patella ㉔ of the left (free-play) leg moves medially/superiorly. Fulcrum left hip joint and pDP left anterior iliac spine ⑧ move caudally/medially in a co-rotational continuing movement (Fig. 221).

Lines: The right/left fold of the groin (XV) deepens, as does the right/left gluteal fold (XIV).

Axes: The line connecting the anterior iliac spines (5) meets the long axis of the right/left thigh (12) in the frontal plane, forming a medial angle of less than 90°.

182

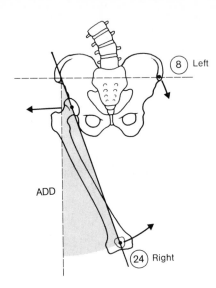

Fig. 220. Adduction at the hip joint effected by dDP ⑧ and pDP ㉔ with displacement of the fulcrum

Remark: In buttressing mobilization (see p. 312) of the hip joint in abduction/adduction, we use the ipsilateral iliac spine as proximal distance point in abduction and the contralateral iliac spine as proximal distance point in adduction, so that the principle of the contrary movements of the distance points is kept clear. When observing movement, it is better to use the contralateral iliac spine for both adduction and abduction because it traces a longer and therefore a more easily recognizable spatial path.

5.3.5 Movements at the Hip Joints About Frontosagittal Axes

Starting position of the examination: Lying prone, the hip joints at zero, the feet hanging out over the end of the bench. The right knee joint is flexed to about 90°. The line connecting the right and left anterior iliac spines is the proximal pointer, the long axis of the lower leg, vertical and sagittotransverse in its zero position, is the distal pointer; the two pointers are at an angle of 90° to each other.

Internal rotation: 90° (zero position), less the dorsolateral angle in the transverse plane of the knee joint, formed where the long axis of the lower leg meets the lateral projection of the line connecting the anterior iliac spines. Movement tolerance 40°–50° (Fig. 222).

Points: pDP right iliac spine ⑧ and the dDP right greater trochanter ㉓ move closer together (or the opposite, for the left leg). Starting with the hip joint at zero, the iliac spine ⑧ remains stationary while ㉓ moves ventrally/medially in relation to it.

Lines: Starting with the hip joint at zero, the fold of the groin (XV) deepens.

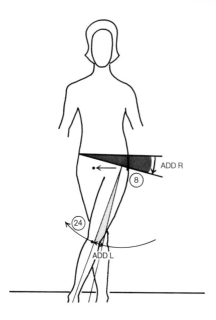

Fig. 221. Adduction at the hip joint in weight bearing

Axes: The line connecting the iliac spines (5) and the flexion/extension axis of the right/left knee joint (23) form an angle in a transverse plane. When, with the hip joint at zero, the moving pointer is the distal one, the popliteal fold of the right/left knee joint (XVI) faces dorsally/laterally. If it is the proximal pointer that moves, the navel ④ faces ventrally/laterally, towards the side of the internally rotated hip joint. pDP right iliac spine has moved laterally/dorsally in relation to dDP right trochanter.

External rotation: 90° (zero position), less the dorsomedial angle in the transverse plane of the knee joint, formed where the long axis of the lower leg meets the line connecting the iliac spines. Movement tolerance 30°–40° (Fig. 222).

Points: pDP right iliac spine ⑧ and dDP right greater trochanter ㉓ move further apart (or the opposite, for the left leg). Starting with the hip joint at zero, the iliac spine ⑧ remains stationary while ㉓ moves dorsally/medially in relation to ⑧.

Lines: The right/left fold of the groin (XV) becomes shallower.

Axes: The line connecting the iliac spines (5) and the flexion/extension axis of the right/left knee joint (23) form an angle in a transverse plane. When, with the hip joint at zero, the moving pointer is the distal one, the popliteal fold of the right/left knee joint (XVI) faces dorsally/medially. When it is the proximal pointer which moves, the navel ④ faces ventrally/laterally, away from the externally rotated hip joint. pDP right iliac spine has moved ventrally/medially in relation to dDP right trochanter.

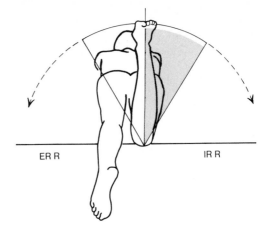

Fig. 222. Internal/external rotation at the right hip joint in prone lying, effected by the distal pointer

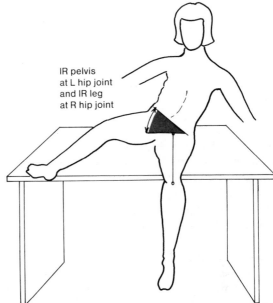

IR pelvis
at L hip joint
and IR leg
at R hip joint

Fig. 223. Internal rotation in sitting with 90 ° flexion: effected at the right hip joint by the distal pointer and at the left hip joint as a continuing movement by the proximal lever

Note

When, in sitting upright, the long axes of the thighs, as axes of rotation of the hip joints, are sagittotransverse, the long axes of the lower legs, as distal rotary pointers, and the line connecting the iliac spines, as proximal pointers, move in frontal planes. When the left lower leg moves in its frontal plane, because of internal rotation at the left hip joint, the co-rotational continuing movement sets the pelvis in motion in its frontal plane, with internal rotation at the right hip joint effected by the proximal pointer. The lumbar spine performs a left concave lateral flexion (Figs. 223, 224).

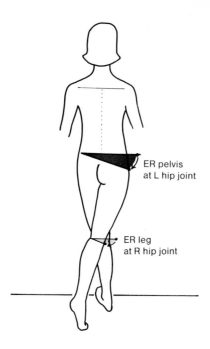

ER pelvis
at L hip joint

ER leg
at R hip joint

Fig. 224. External rotation in standing: effected at the right hip joint by the distal pointer and at the left hip joint as a corotational continuing movement by the proximal pointer

Transverse abduction with the long axis of the thigh sagittotransverse (zero position): the medial angle formed by the line connecting the iliac spines and the long axis of the right/left thigh in the transverse plane of the right/left hip joint, less the 90° of the zero position. Movement tolerance 60°.

Points: dDP right patella ㉔ moves away from the pDP left iliac spine ⑧, ㉔ moving to the right laterally and dorsally (or, for the left leg, the opposite occurs).

Lines: The right/left fold of the groin (XV) becomes shallower.

Axes: The line connecting the iliac spines (5) meets the long axis of the thigh (12) at an angle greater than 90° in a transverse plane. If it is the proximal lever that moves, pDP left iliac spine ⑧ moves laterally/dorsally.

Transverse adduction with the long axis of the thigh sagittotransverse (zero position): the 90° of the zero position, less the medial angle in the transverse plane of the right/left hip joint formed by the line connecting the iliac spines and the long axis of the right/left thigh. Movement tolerance 30°.

Points: dDP right patella ㉔ and pDP left iliac spine ⑧ move closer together, ㉔ moving medially/dorsally (the opposite happens for transverse adduction at the left hip joint).

Lines: The right/left fold of the groin (XV) deepens.

Axes: The line connecting the iliac spines (5) meets the long axis of the thigh at an angle of less than 90° in a transverse plane. When it is the proximal lever that moves, pDP left iliac spine ⑧ moves ventrally/medially.

> **Note**
> The continuing movement pertaining to transverse abduction and adduction is rotation in the thoracic spine.

5.3.6 Movements at the Knee Joints (Figs. 225, 226)

> **Flexion:** 180° (between the long axes of the thigh and of the lower leg in zero position), less the dorsal angle formed between these two axes. Movement tolerance 120°.

Points: pDP greater trochanter ㉓ and dDP lateral malleolus ㉖ move closer together. For observation from a distance and for the purpose of instruction, the ischial tuberosity ㉑ can be used as proximal distance point and the heel ㉘ as the distal distance point.

Lines: The popliteal fold (XVI) deepens.

Axes: The long axis of the thigh (12) and the long axis of the lower leg (13) meet at an angle of less than 180°; the vertex of this angle never points dorsally.

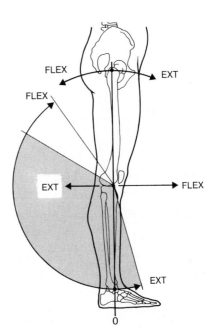

Fig. 225. Flexion/extension at the knee joints

187

Points: pDP greater trochanter ㉓ and dDP lateral malleolus ㉖ move towards each other ventrally of the flexion/extension axis of the knee joint.

Lines: The popliteal fold (XVI) becomes shallower.

Axes: The long axes of the thigh (12) and of the lower leg (13) form an angle of less than 180°, the vertex of which never points ventrally.

Remark: When the leg is weight bearing, the usual way in which extension and flexion at the knee joint take place is through dorsal (in extension) and ventral (in flexion) gliding of the fulcrum.

Note

Rotary movement excursions at the knee joint are normally only possible when the joint is flexed; the long axis of the lower leg is the axis of rotation. With 90° flexion at the knee joint, the long axis of the thigh is the proximal pointer and the transverse axis through the head of the tibia or the anatomical long axis of the foot is the distal pointer (Fig. 226).

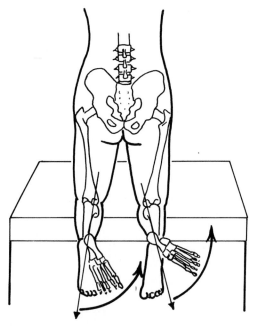

IR about long axis
of L lower leg

ER about long axis
of R lower leg

Fig. 226. Internal/external rotation at the knee joint about the long axis of the lower leg (kneeling)

188

External rotation in the 90°-flexed knee joint from the zero position (90° angle between the transverse axis of the tibial head and the long axis of the thigh): the 90° of the zero position less the lateral angle between the transverse axis of the tibia and the long axis of the thigh. Movement tolerance 40°.

Points: dDP right/left tibial tuberosity ㊹ moves dorsally/laterally in relation to pDP right/left lateral femoral epicondyle ㊽, or pDP right/left greater trochanter ㉓ moves ventrally/laterally in relation to dDP right/left tibial tuberosity ㊹ in the transverse plane of the right/left knee joint.

Lines: The anterior border of the right/left tibia (XVIII) moves laterally. The lateral border of the right/left foot (XVII) moves outwards (diverges).

Axes: The functional long axis of the right/left foot (42) and the long axis of the right/left thigh (12) form a lateral angle of less than 180°. The difference between this and the zero position (180°) equals the amount of external rotation.

Remark: If the external rotation at the knee joint is effected by the distal pointer, dDP tibial tuberosity moves laterally/dorsally (this can be observed in a leg in free play or parking function). If the external rotation is effected by the proximal pointer, pDP greater trochanter moves laterally/ventrally (this can be observed in a weight-bearing leg (Fig. 226).

Internal rotation in the 90°-flexed knee joint from the zero position (90° angle between the transverse axis of the tibial head and the long axis of the thigh): the 90° of the zero position less the medial angle between the transverse axis of the tibia and the long axis of the thigh. Movement tolerance 10°.

Points: dDP right/left tibial tuberosity ㊹ moves dorsally/medially in relation to pDP right/left medial femoral epicondyle ㊾, or pDP greater trochanter ㉓ moves ventrally/medially in relation to dDP right/left tibial tuberosity ㊹.

Lines: The ventral border of the right/left tibia (XVIII) moves medially. The lateral border of the right/left foot (XVII) moves inwards (converges).

Axes: The functional long axis of the right/left foot (42) and the long axis of the right/left thigh (12) form a medial angle of less than 180°. The difference between this and the zero position (180°) is the amount of internal rotation.

Remark: The same remark as for external rotation, except that for 'lateral' read 'medial'.

5.3.7 Movements at the Talocrural Joints (Fig. 227)

Starting position: Standing upright.

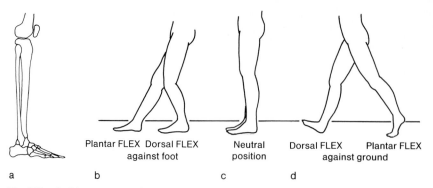

Plantar FLEX Dorsal FLEX   Neutral   Dorsal FLEX   Plantar FLEX
      against foot   position   against ground

a   b   c   d

Fig. 227 a–d. Movements at the talocrural joint. **a** Skeleton; **b** plantar flexion and dorsiflexion of the lower leg against the foot; **c** neutral position; **d** plantar flexion and dorsiflexion of the foot against the ground

Dorsiflexion: The 90° of the zero position, less the anterior angle between the long axis of the lower leg and the lateral border of the foot. Movement tolerance 30°.

Points: dDP metatarsophalangeal joint of the right/left large toe ㉙ and pDP right/left tibial tuberosity ㊹ move closer together.

Lines: The right/left Achilles tendon (XIX) becomes longer.

Axes: The ventral angle between the anatomical long axis of the right/left foot (11) and the long axis of the right/left lower leg (13) becomes smaller. The distance between the flexion/extension axis of the right/left knee joint (23) and the base of the transverse arch of the right/left foot (20) decreases.

Plantar flexion: The ventral angle between the long axis of the lower leg and the lateral border of the foot, less the 90° of zero position. Movement tolerance 50°.

Points: dDP metatarsophalangeal joint of the right/left large toe ㉙ and pDP right/left tibial tuberosity ㊹ move further apart.

Lines: The right/left Achilles tendon (XIX) becomes shorter.

Axes: The ventral angle between the anatomical long axis of the right/left foot (11) and the long axis of the right/left lower leg (13) increases. The distance between the flexion/extension axis of the right/left knee joint (23) and the base of the transverse arch of the right/left foot (20) increases.

5.3.8 Movements at the Subtalar Joints (Fig. 228)

Starting position: Sitting upright with the feet the width of the pelvis apart, in line with the hip joints, with 90° flexion of the knee joints; the soles of the feet are in contact with the floor and the functional long axes of the feet point forwards.

190

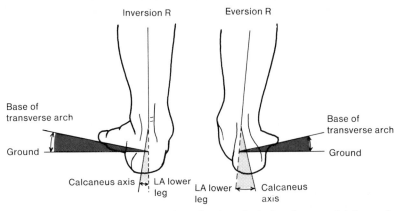

Fig. 228. Movements at the right talocrural joint (inversion/eversion): dorsal aspect. *LA*, long axis

Eversion: The lateral angle formed by the surface of the heel and the transverse arch of the foot (without extra pronation) with the floor when the lateral border of the foot is raised while the fibular head remains stationary. Movement tolerance 15°.

Points: pDP right/left lateral malleolus ㉖ and dDP tubercle of the right/left lateral calcaneal tuberosity ㊿ move closer together; the right/left fibular head remains stationary.

Lines: The skin creases laterally below the right/left lateral malleolus.

Axes: The top of the right/left calcaneal tuberosity (28) and the long axis of the right/left lower leg (13) form an angle with its vertex pointing medially.

Inversion: The medial angle formed by the surface of the heel and the transverse arch of the foot (without extra supination) with the floor when the medial border of the foot is raised. Movement tolerance 35°.

Points: pDP right/left medial malleolus ㉗ and dDP right/left tubercle (medial process) of the calcaneal tuberosity ㊶ move close together; the right/left fibular head remains stationary.

Lines: The skin creases medially below the right/left medial malleolus.

Axes: The top of the right/left calcaneal tuberosity (28) and the long axis of the right/left lower leg (13) form an angle with its vertex pointing laterally.

Remark: Eversion and inversion are movements of the hindfoot which contribute to the formation of the longitudinal arch of the foot.

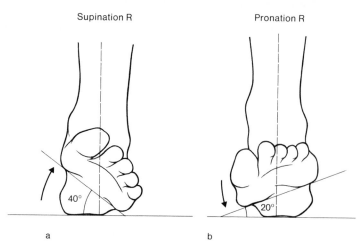

Supination R   Pronation R

40°   20°

a   b

Fig. 229 a, b. Movements at the Chopart and Lisfranc articulations (supination/pronation; ventral aspect)

5.3.9 Movements at the Chopart and Lisfranc Joints (Fig. 229)

Supination: Elevation of the medial border of the foot about an axis which passes through the calcaneus and the third toe with the hindfoot firmly on the ground; the twist of the forefoot is reduced. Movement tolerance 40°.

Points: dDP head of the right/left fifth metatarsal ③⓪ and pDP right/left lateral malleolus ②⓪ move away from each other.

Lines: The longitudinal arch of the right/left foot (XX) is flattened.

Axes: The base of the transverse arch of the right/left foot (20) forms an angle with the floor, the vertex of which points laterally. The top of the right/left calcaneal tuberosity (28) does not move out of the long axis of the right/left lower leg (13) (Fig. 229 a).

Pronation: Elevation of the lateral border of the foot about an axis which passes through the calcaneus and the third toe with the hindfoot firmly on the ground; the twist of the forefoot is increased. Movement tolerance 20°.

Points: dDP head of the right/left fifth metatarsal ③⓪ and pDP right/left lateral malleolus ②⑥ move closer together.

Lines: The longitudinal arch of the right/left foot (XX) is accentuated.

Axes: The base of the transverse arch of the right/left foot (20) forms an angle with the floor, the vertex of which points medially. The top of the right/left calcaneal tuberosity (28) does not move out of the long axis of the lower leg (13) (Fig. 229 b).

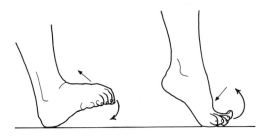

Fig. 230. Flexion/extension at the phalangeal joints

Remark: The longitudinal arch of the foot is a structure which can be accentuated or flattened by buttressing activities. Inversion of the hindfoot and pronation of the fore-foot accentuate the longitudinal arch; eversion of the hindfoot and supination of the forefoot reduce or eliminate it. These mechanisms help the soles of the feet to adjust to sideways shifting of weight and sloping or uneven ground.

5.3.10 Movements at the Toe Joints

The metatarsophalangeal joints are ellipsoid. Abduction (spreading) and adduction (closing) relate to the anatomical long axis of the foot. The movement tolerance in flexion is 40°, in extension 70°, with the large toe possessing the greatest range of flexion.

The proximal interphalangeal joints of the toes possess a movement tolerance of 35° flexion (the large toe 80°); there is no extension. The distal interphalangeal joints possess a movement tolerance of 60° flexion and 30° extension (according to Debrunner, 1971).

Because of passive insufficiency of the extensors the foot must be dorsiflexed at the talocrural joint to achieve the total range of 60° flexion at the phalangeal joints.

Because of passive insufficiency of the flexors, the foot must be plantarflexed at the talocrural joint to achieve the total range of extension at the phalangeal joints (Fig. 230).

5.4 Points, Lines and Axes of BS Arms

Normal appearance in the upright stance in the zero position: BS pelvis, BS thorax and BS head are aligned in the long axis of the body, which is vertical. The feet are on the floor, the width of the pelvis apart. The thoracic spine is dynamically stabilized at zero. The pelvis is potentially mobile at the hip joints and the joints of the lumbar spine; the joints of the cervical spine and the atlanto-occipital and atlanto-axial joints are also potentially mobile. In this posture, the shoulder girdle, positioned over the thorax, is parked and the arms in free play, hanging from the shoulder girdle.

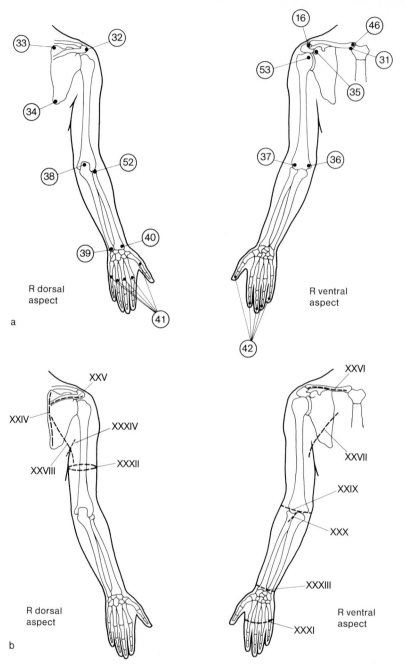

Fig. 231a–c. Points, lines and axes on BS arms. **a** *Points:* ⑯ Right/left acromioclavicular joint; ㉛ right/left sternoclavicular joint; ㉜ right/left acromion; ㉝ right/left superior angle of scapula; ㉞ right/left inferior angle of scapula; ㉟ right/left coracoid process; ㊱ right/left medial humeral epicondyle; ㊲ right/left lateral humeral epicondyle; ㊳ right/left olecranon; ㊴ right/left ulnar styloid process; ㊵ right/left radial styloid process; ㊶ right/left metacarpophalangeal joints of fingers I–V; ㊷ right/left finger tips I–V; ㊻ right/left sternal end of clavicle; ㊵ right/left head of radius; ㊾ right/left head of humerus. **b** *Lines:* XXIV right/left medial border of scapula; XXV right/left scapular spine; XXVI right/left clavicle; XXVII right/left ventral border of axilla; XXVIII right/left dorsal border of axilla; XXIX right/left fold of elbow joint; XXX right/left aponeurosis of biceps; XXXI right/left transverse palmar line of the hand (metacarpophalangeal

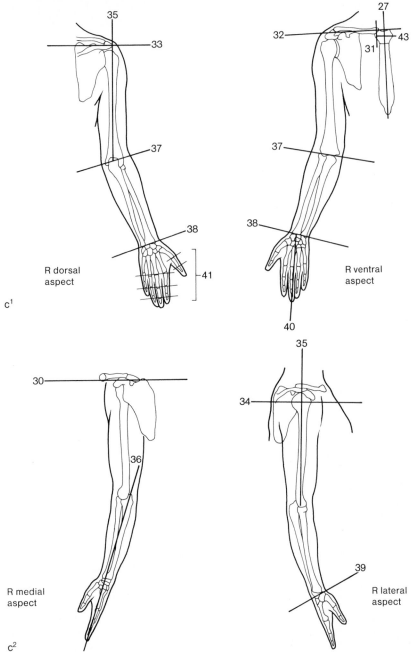

c¹   R dorsal aspect   R ventral aspect

c²   R medial aspect   R lateral aspect

joints of the fingers); XXXII right/left circumference of mid upper arm; XXXIII right/left fold of wrist; XXXIV right/left fold of axilla. **c** *Axes:* 27 long axis of sternum; 30 right/left sagitto-transverse axis of sternoclavicular joint (cranio-/caudoduction of the shoulder); 31 right/left frontosagittal axis of sternoclavicular joint (ventro-/dorsiduction of the shoulder); 32 right/left long axis of clavicle (axis of rotation); 33 right/left flexion/extension axis of shoulder joint; 34 right/left abduction/adduction axis shoulder joint; 35 right/left long axis of upper arm (external/internal rotation axis of shoulder joint); 36 right/left long axis of forearm (pronation/supination axis); 37 right/left flexion/extension axis of elbow joint; 38 right/left flexion/extension axis of wrist; 40 right/left long axis of hand from the middle of the wrist to the middle finger; 41 right/left flexion/extension axis of finger joints; 43 transverse axis of the manubrium sterni

The joints of BS arms, movement excursions of which can be functionally measured by measuring the changes in position and shape of points, lines and axes, are as follows (Fig. 231):

The sternoclavicular, acromioclavicular and humeroscapular joints, the elbow joints, the wrist joints (radiocarpal and intercarpal) and those of the fingers.

When discussing the joints of BS arms we differentiate between proximal and distal distance points, using the abbreviations pDP and dDP.

BS arms articulates with BS thorax at the sternoclavicular joints.

When dealing with the switch points of the closely associated sternoclavicular, acromioclavicular and humeroscapular joints, we are looking at functional units and must analyse them as such when observing movement sequences. It is important to examine the function of these three levels of movement independently of each other, seeing how the clavicle moves against the sternum, the scapula against the clavicle, and the humerus against the scapula.

The sternoclavicular joint really provides only a very tenuous link between BS arms and the remainder of the body. The advantage of this tenuous link is that it enables each shoulder joint to alter its position in relation to the thorax independently, ensuring always the best possible position for the hands in the performance of their various activities. However, with this considerable degree of mobility, how is it that the connection between BS arms and the remainder of the body is in fact quite sturdy?

The functional mechanism of the three levels of movement clavicle/sternum, scapula/clavicle and humerus/scapula can be compared with the action of a pair of pincers. The scapula and clavicle form the opening, the jaws of the pincers. The pivot of the jaws is where the scapula articulates with the clavicle, close to the humeroscapular joint (Fig. 232). Looking more closely at the jaws of the pincers, we see that they are always open, because the thorax is held between them. The points where the pincers grip ventrally, the fragile attachments of the clavicles to the sternum, are the fixed points in relation to the rest of the body. For this reason, this level of movement needs no direct bridging musculature. Since the thorax is permanently positioned between the jaws of the pincers, the mobile lower jaw, the scapula, having no proximal articulation with any other part of the skeleton, must be connected in a very special way with the remainder of the body. This connection is achieved by slings of muscles capable of very finely differentiated activities; their insertions stretch not only to the arms but beyond and across to the thorax and the pelvis. They maintain the position of the shoulder girdle and the arms over the thorax or move the latter beneath the shoulder girdle.

This arrangement works extremely well, both for movement and for stabilization, as long as the shape of the thorax is normal. If it is abnormal, the connection between the shoulder girdle and the thorax is altered (see p. 159).

Remark: We differentiate between the proximal joints and the distal joint at the shoulder. The proximal joints are the sternoclavicular joint and the acromioclavicular joint, which necessarily moves with it; the distal joint is the humeroscapular joint, the one usually meant when speaking of 'the shoulder joint'. It is important to differentiate between the functions of the proximal joints and the distal joint when treating them.

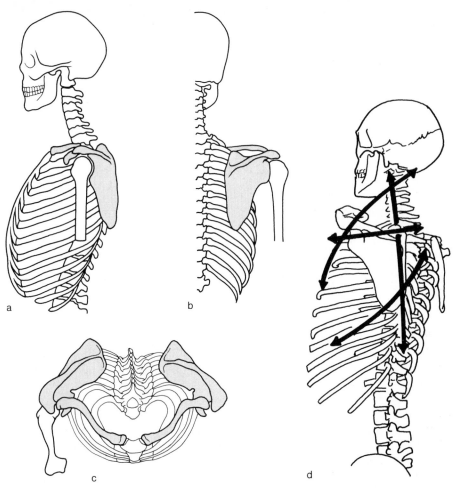

Fig. 232 a-d. The jaws of the pincers. **a** Lateral aspect, **b** dorsal aspect, **c** cranial aspect, **d** the sling-like system of muscles which serve to move the jaws of the pincers on the thorax or attach the thorax to the pincers (Hoeppke 1971)

5.4.1 Movements of the Clavicle at the Sternoclavicular and Acromioclavicular Joints (Proximal Shoulder Joints)

Cranioduction of the shoulder joint: The clavicle moves about the sagittotransverse axis while the acromioclavicular joint moves cranially/medially in a frontal plane. The angle formed between the clavicle and the long axis of the sternum in the zero position (slightly more than 90°) increases. Movement tolerance is 60°. The angle between the clavicle and the scapula – the 'pincer jaws' – decreases (Fig. 233).

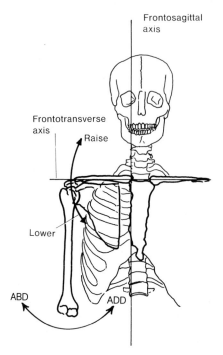

Fig. 233. Cranioduction and caudoduction of the shoulder joint

Points: dDP right/left acromioclavicular joint ⑯ moves cranially/medially in relation to pDP suprasternal notch ③. dDP upper angle of the right/left scapula ㉝ moves towards pDP C7 ⑤. dDP lower angle of the right/left scapula ㉞ moves away from the line connecting the spinous processes of T1–T12 (I).

Lines: The ventral and dorsal borders of the right/left axilla (XXVII and XXVIII) become longer. The right/left lateral line of the neck between the ear lobe and the right/left acromioclavicular joint (IX) shortens, if the head is aligned in the long axis of the body.

Axes: The caudal angle between the long axis of the sternum (27) and the long axis of the right/left clavicle (32) increases.

Caudoduction of the shoulder joint: The clavicle moves about the sagittotransverse axis while the acromioclavicular joint moves caudally/laterally in a frontal plane. The angle formed between the clavicle and the long axis of the sternum decreases. Movement tolerance is 5°. The angle between the clavicle and the scapula – the 'pincer jaws' – increases (Fig. 233).

Points: dDP right/left acromioclavicular joint ⑯ moves caudally/laterally in relation to pDP suprasternal notch ③. dDP upper angle of the right/left scapula ㉝ moves

198

away from pDP C7 ⑤. dDP lower angle of the right/left scapula ㉞ moves towards the line connecting the spinous processes of T1–T12 (I).

Lines: The ventral and dorsal borders of the right/left axilla (XXVII and XXVIII) become shorter. The lateral line of the neck between the right/left ear lobe and the right/left acromioclavicular joint (IX) becomes longer.

Axes: The caudal angle between the long axis of the sternum (27) and the long axis of the clavicle (32) decreases.

Ventroduction of the shoulder joint (protraction): The clavicle moves about the frontosagittal axis while the acromioclavicular joint moves ventrally/medially in a transverse plane. The ventral angle between the clavicle and the transverse axis of the manubrium sterni (180° in zero position) decreases. Movement tolerance is 45°. The angle between the clavicle and the scapula – the 'pincer jaws' – decreases (Fig. 234).

Points: dDP right/left acromioclavicular joint ⑯ moves ventrally/medially in relation to pDP suprasternal notch ③. dDPs upper ㉝ and lower ㉞ angles of the right/left scapula move further away from the line connecting the spinous processes of T1–T12 (I).

Lines: The ventral border of the right/left axilla (XXVII) becomes shorter while the dorsal border (XXVIII) becomes longer.

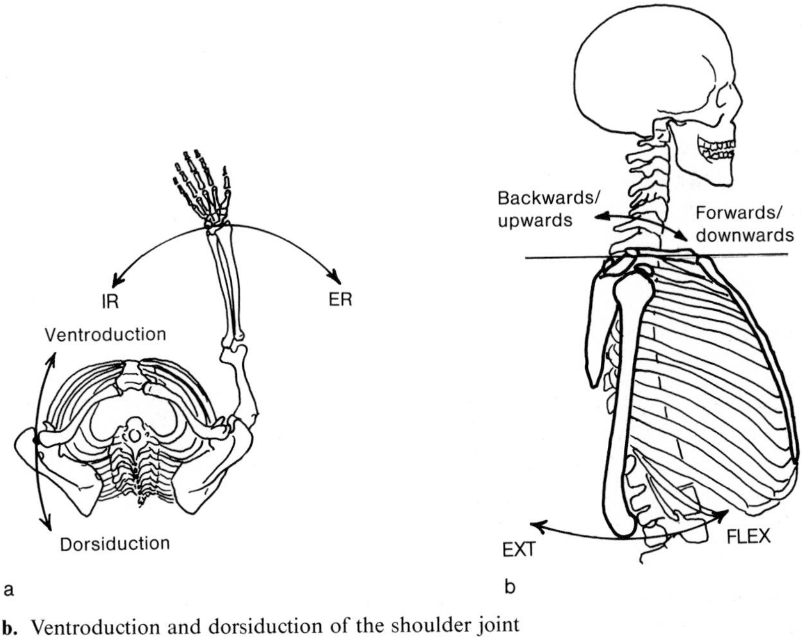

Fig. 234 a, b. Ventroduction and dorsiduction of the shoulder joint

Axes: The ventral angle between the transverse axis of the manubrium sterni (43) and the long axis of the right/left clavicle (32) decreases to less than 180°.

Dorsiduction of the shoulder joint (retraction): The clavicle moves about the fronto-sagittal axis while the acromioclavicular joint moves dorsally/medially in a transverse plane. The ventral angle between the clavicle and the transverse axis of the manubrium sterni (180° in zero position) increases. Movement tolerance is 20°. The angle between the clavicle and the scapula – the 'pincer jaws' – increases (Fig. 234).

Points: dDP right/left acromioclavicular joint ⑯ moves dorsally/medially in relation to pDP suprasternal notch ③. dDPs upper ㉝ and lower ㉞ angles of the right/left scapula move towards the line connecting the spinous processes of T1–T12 (I).

Lines: The ventral border of the right/left axilla (XXVII) becomes longer while the dorsal border (XXVIII) becomes shorter. The medial border of the right/left scapula (XXIV) and the line connecting the spinous processes of T1–T12 (I) move closer together.

Axes: The ventral angle between the long axis of the right/left clavicle (32) and the transverse axis of the manubrium sterni (43) increases to more than 180°.

Ventral rotation of the clavicle: The clavicle rotates about its own long axis, which is almost frontotransverse, while the acromioclavicular joint moves ventrally/caudally in a sagittal plane. The clavicle rotates ventrally about its own long axis at the sternoclavicular joint. The angle between the clavicle and the scapula – the 'pincer jaws' – decreases (Fig. 235).

Points: dDP right/left acromioclavicular joint ⑯ moves ventrally/caudally in relation to pDP suprasternal notch ③. dDP lower angle of the right/left scapula ㉞ moves away from the posterior wall of the thorax.

Lines: The medial border of the right/left scapula (XXIV) glides cranially.

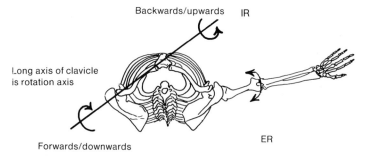

Backwards/upwards   IR

Long axis of clavicle
is rotation axis

Forwards/downwards   ER

Fig. 235. Moving the shoulder joint forwards/downwards and backwards/upwards: ventral rotation of the clavicle

Axes: The ventral/caudal angle formed by the long axis of the sternum (27) and the sagittotransverse abduction/adduction axis of the right/left shoulder joint (34), which in zero position is 90°, decreases.

Dorsal rotation of the clavicle: The clavicle rotates about its own long axis, which is almost frontotransverse, while the acromioclavicular joint moves dorsally/cranially in a sagittal plane. The clavicle rotates dorsally about its own long axis at the sternoclavicular joint. The angle between the clavicle and the scapula – the 'pincer jaws' – increases (Fig. 235).

Points: dDP right/left acromioclavicular joint ⑯ moves dorsally/cranially in relation to pDP suprasternal notch ③. dDP lower angle of the right/left scapula ㉞ moves closer to the posterior wall of the thorax.

Lines: The medial border of the right/left scapula (XXIV) glides caudally.

Axes: The ventral/caudal angle formed by the long axis of the sternum (27) and the sagittotransverse abduction/adduction axis of the right/left shoulder joint (34), which in the zero position measures 90°, increases.

5.4.2 Movements at the Humeroscapular Joints (Distal Shoulder Joints)

Abduction: Movements of the humerus and (possibly but not necessarily) the shoulder girdle about the sagittotransverse axis in frontal planes. The olecranon moves laterally/cranially, or the lower angle of the scapula moves medially. Movement tolerance 90° (Fig. 236).

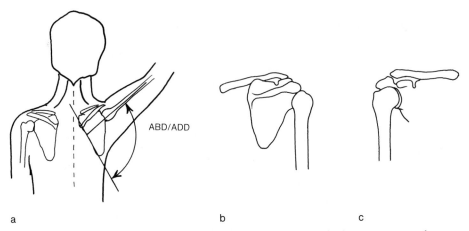

a   b   c

Fig. 236. a Abduction/adduction at the right shoulder joint and continuing movement at the sternoclavicular joint. **b,c** Skeleton of right shoulder joint: **b** dorsal, **c** ventral aspect

Points: pDP lower angle of the right/left scapula ㉞ and dDP right/left olecranon ㊳ move further apart in a frontal plane (the plane of the scapula, to be exact).

Lines: The fold of the right/left axilla (XXXIV) becomes shallower.

Axes: The long axis of the right/left upper arm (35) forms in the frontal (scapular) plane a caudal angle of not more than 90° with the medial border of the right/left scapula (XXIV). The long axis of the right/left clavicle (32) forms an angle of not more than 180° with the long axis of the upper arm (35).

Adduction: Movement of the humerus and (possibly but not necessarily) the shoulder girdle about the sagittotransverse axis in frontal planes. The olecranon moves medially/cranially, or the lower angle of the scapula moves laterally/cranially. Movement tolerance 20°–40° (Fig. 236).

Points: pDP lower angle of the right/left scapula ㉞ and dDP right/left olecranon ㊳ move towards each other in a frontal plane (the scapular plane).

Lines: The fold of the right/left axilla (XXXIV) becomes deeper.

Axes: The long axis of the right/left upper arm (35) forms a caudal/medial angle of less than 90° with the long axis of the right/left clavicle (32).

Flexion: Movement of the humerus and (possibly but not necessarily) the shoulder girdle about the frontotransverse axis in sagittal planes. The olecranon moves ventrally/cranially or the lower angle of the scapula moves dorsally/cranially. Movement tolerance 110° (Fig. 237).

Points: pDP lower angle of the right/left scapula ㉞ and dDP right/left olecranon ㊳ move away from each other in a sagittal plane; dDP right/left olecranon moves ventrally/cranially.

Lines: The fold of the right/left axilla (XXXIV) becomes shallower ventrally and deeper dorsally.

Axes: The long axis of the right/left upper arm (35) and the medial border of the right/left scapula (XXIV) form a ventral/caudal angle of not more than 110° in the sagittal plane.

Extension: Movement of the humerus and (possibly but not necessarily) the shoulder girdle about the frontotransverse axis in sagittal planes. The olecranon moves dorsally/cranially or the lower angle of the scapula moves ventrally. Movement tolerance 40°.

Points: pDP lower angle of the right/left scapula ㉞ and dDP right/left olecranon ㊳ move away from each other in a sagittal plane; dDP the right/left olecranon moves ventrally/cranially.

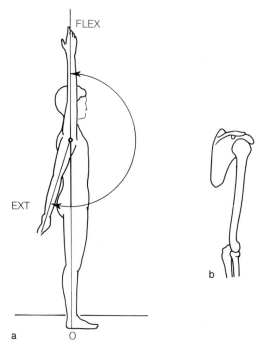

Fig. 237. a Flexion/extension at the right shoulder joint and continuing movement at the sterno-clavicular joint. **b** Skeleton of right shoulder joint, lateral aspect

Lines: The fold of the right/left axilla (XXXIV) becomes shallower ventrally and deeper dorsally.

Axes: The long axis of the right/left upper arm (35) and the medial border of the right/left scapula (XXIV) form a dorsal/caudal angle of not more than 40°.

Transverse flexion: Movement of the humerus and (possibly but not necessarily) the shoulder girdle about the frontosagittal axis in transverse planes from the zero position, with the long axis of the upper arm frontotransverse. Movement tolerance is 90° and the olecranon moves ventrally/medially (Fig. 238).

Points: dDP right/left olecranon ㊳ and pDP right/left sternal end of the clavicle ㊻ move towards each other in a transverse plane. pDPs upper/lower angle of the right/left scapula ㉝/㉞ and right/left olecranon ㊳ move towards each other in a transverse plane.

Lines: The ventral border of the right/left axilla (XXIV) becomes shorter, the dorsal border (XXVIII) becomes longer. When the shoulder girdle is drawn into the continuing movement the medial border of the right/left scapula (XXIV) moves away from the line connecting the spinous processes of T1–T12 (I).

203

Axes: The long axis of the right/left clavicle (32) and the long axis of the right/left upper arm (35) meet at an angle ventrally. The difference between this angle and the 180° of the zero position is the extent of the transverse flexion.

Transverse extension: Movement of the humerus and (possibly but not necessarily) the shoulder girdle about the frontosagittal axis in transverse planes from the zero position, with the long axis of the upper arm frontotransverse. Movement tolerance is 40° and the olecranon moves dorsally/medially (Fig. 238).

Points: dDP right/left olecranon ㊳ and pDP upper angle of the right/left scapula ㉝ move towards each other in a transverse plane.

Lines: The dorsal border of the right/left axilla (XXVIII) becomes shorter, the medial border (XXVII) becomes longer. When the shoulder girdle is drawn into the continuing movement, the medial border of the right/left scapula (XXIV) moves towards the line connecting the spinous processes of T1–T12 (I).

Axes: The long axis of the right/left clavicle (32) and the long axis of the right/left upper arm (35) meet at an angle ventrally. This angle, less the 180° of the zero position, represents the extent of the transverse extension.

Internal rotation: Medial rotation of the humerus about its own frontosagittal long axis; the shoulder girdle may also rotate about the same axis. The proximal pointer (the long axis of the clavicle) and the distal pointer (the flexion/extension axis of the elbow joint or, with the elbow flexed at 90°, the long axis of the forearm) move in transverse planes. Movement tolerance is 95° (Fig. 239a).

Points: In relation to pDP right/left acromion ㉜, dDP right/left lateral humeral epicondyle ㊲ moves ventrally/medially and dDP right/left medial humeral epicondyle ㊱ moves dorsally/laterally.

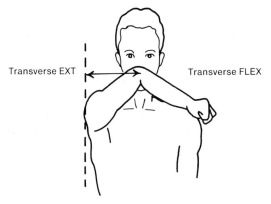

Fig. 238. Transverse flexion/extension at the right shoulder joint and continuing movement at the sternoclavicular joint

204

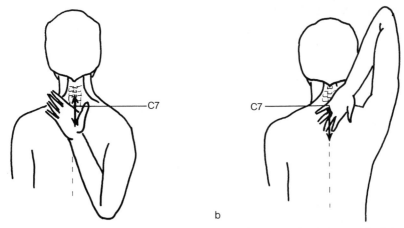

Fig. 239 a, b. Maximum internal rotation (**a**) and external rotation (**b**) at the right shoulder joint with continuing movement at the sternoclavicular joint

Lines. The crease of the right/left elbow joint (XXIX) faces more medially.

Axes: When the right/left humeroscapular joint is in the zero position, the flexion/extension axis of the elbow joint (37), in relation to the midfrontal plane, is slightly medially rotated. At maximum internal rotation it is almost sagittotransverse and forms an angle of about 90° with the long axis of the clavicle.

External rotation: Lateral rotation of the humerus about its own frontosagittal long axis; the shoulder girdle may also rotate about the same axis. The proximal pointer (the long axis of the clavicle) and the distal pointer (the flexion/extension axis of the elbow joint or, with the elbow flexed at 90°, the long axis of the forearm) move in transverse planes. Movement tolerance is 40°–60° (Fig. 239 b).

Points: In relation to pDP right/left acromion ㉜, dDP right/left lateral humeral epicondyle ㊲ moves laterally/dorsally and dDP right/left medial humeral epicondyle ㊱ moves medially/ventrally.

Lines: The crease of the right/left elbow joint (XXIX) faces more laterally.

Axes: When the right/left humeroscapular joint is in the zero position, the flexion/extension axis of the right/left elbow joint (37), in relation to the midfrontal plane, is slightly medially rotated. At maximum external rotation it forms a dorsal angle of 125°–145° with the long axis of the right/left clavicle.

Remark: The sternoclavicular joint is normally drawn into distally arising movements at the humeroscapular joint, in the form of a co-rotational continuing movement. Resulting from this, the following combinations can occur:

Humeroscapular joint	Sternoclavicular joint
Abduction	Shoulder joint (DP acromion) raised in the frontal plane
Adduction	Shoulder joint (DP acromion) lowered in the frontal plane
Flexion	Dorsal/cranial movement of the acromioclavicular joint in the sagittal plane
Extension	Ventral/caudal movement of the acromioclavicular joint in the sagittal plane
Transverse flexion or internal rotation	Ventral movement of the shoulder joint (DP acromion) in the transverse plane
Transverse extension or external rotation	Dorsal movement of the shoulder joint (DP acromion) in the transverse plane

5.4.3 Movements at the Elbow Joints and of the Forearm

Flexion: 180° (between the long axes of the upper arm and the forearm, i.e. zero position), less the ventral angle between the axes of the upper arm and the forearm. Movement tolerance is 150° (Fig. 240).

Points: dDPs styloid processes of the right/left ulna ③⑨ and radius ④⓪ and pDPs ventral aspect of the right/left humeral head and right/left acromion ③② move towards each other.

Lines: The crease on the flexion side of the right/left elbow joint (XXIX) deepens.

Axes: The long axis of the right/left upper arm (35) and the long axis of the right/left forearm (36) form a ventral angle of not smaller than 30°.

Extension: 180° (zero position), less the ventral angle between the long axes of the upper arm and the forearm. Movement tolerance is 10° (Fig. 240).

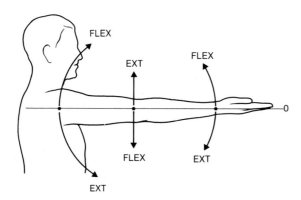

Fig. 240. Flexion/extension at the right elbow joint

Points: dDPs styloid processes of the right/left ulna ㉟ and radius and pDPs ventral aspect of the right/left humeral head and right/left acromion ㉜ move away from each other from the 150° position of maximum flexion to the zero position and move towards each other again in the 10° extension.

Lines: The crease of the right/left elbow joint (XXIX) becomes shallower.

Axes: The long axis of the right/left upper arm (35) and the long axis of the right/left forearm (36) form a dorsal angle of between 180° (zero position) and 170° (extension to end-stop).

Remark: There is a physiological valgus angulation between the long axes of the upper arm and the forearm in the zero position.

Pronation of the forearm: Internal rotating of the hand about the long axis of the forearm. In the zero position, the flexion/extension axis of the elbow joint (the proximal pointer) is almost at right angles to that of the radiocarpal joint (the distal pointer). When the forearm is pronated to end-stop they are almost parallel. Movement tolerance is 90° (Fig. 241).

Points: dDP right/left radial styloid process ㊵ moves medially in relation to pDP right/left lateral humeral epicondyle ㊲.

Lines: The crease of the right/left radiocarpal joint (XXXIII) and that of the right/left elbow joint (XXIX) face in opposite directions.

Axes: The flexion/extension axis of the right/left radiocarpal joint (38) and that of the right/left elbow joint (37) are almost parallel.

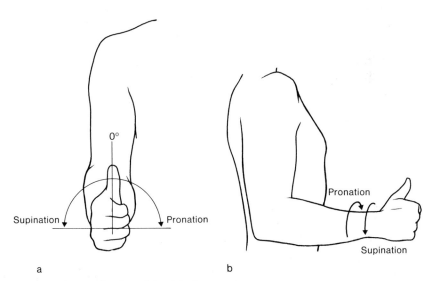

a   b

Fig. 241 a, b. Pronation/supination of the right forearm

Supination of the forearm: External rotation of the hand about the long axis of the forearm. In the zero position the flexion/extension axis of the elbow joint (the proximal pointer) is almost at right angles to that of the radiocarpal joint (the distal pointer). When the forearm is supinated to end-stop they are almost parallel. Movement tolerance is 90° (Fig. 241).

Points: dDP right/left radial styloid process ④⓪ moves laterally in relation to pDP right/left medial humeral epicondyle ㉚.

Lines: The crease of the right/left radiocarpal joint (XXXIII) and that of the right/left elbow joint (XXIX) face the same direction.

Axes: The flexion/extension axis of the right/left radiocarpal joint (38) and that of the right/left elbow joint (37) are almost parallel.

5.4.4 Movements at the Wrist Joints

We differentiate between the proximal wrist joint, the radiocarpal joint, and the distal wrist joints, the intercarpal joints. At these joints, the hand moves about two transverse axes, almost at right angles to each other.

Flexion: 180° (between the long axes of the forearm and the hand, zero position), less the palmar angle formed between the long axes of the forearm and the hand. Movement tolerance is 85°, 50° of which belongs to the proximal wrist joint (Fig. 242).

Points: dDP metacarpophalangeal joints of the right/left fingers II–V ④① and pDP right/left olecranon ㊳ move towards each other (when the arm is in pronation).

Lines: The creases of the right/left radiocarpal joint (XXXIII) become deeper.

Axes: The long axis of the right/left hand (40) and the long axis of the right/left forearm (36) form a palmar angle of less than 180°.

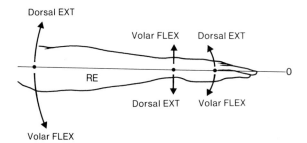

Fig. 242. Flexion/extension at the joints of the hand

208

Extension: 180° (between the long axes of the forearm and the hand, zero position), less the dorsal angle formed between the long axes of the forearm and the hand. Movement tolerance is 85°, 35° of which belongs to the proximal wrist joint (Fig. 242).

Points: dDP metacarpophalangeal joints of the right/left fingers II–V ④① and pDP right/left radial hand ⑤② move towards each other.

Lines: The creases of the right/left radiocarpal joint (XXXIII) become shallower.

Axes: The long axes of the right/left hand (40) and the long axis of the right/left forearm (36) form a dorsal angle of less than 180°.

Ulnar abduction: 180° (between the long axes of the forearm and the hand, zero position), less the ulnar angle formed between the long axes of the forearm and the hand. Movement tolerance is 30°–40° (Fig. 243).

Points: In pronation dDP metacarpophalangeal joint of the right/left finger V ④① moves towards PDP right/left radial head ⑤②; in supination it moves towards pDP right/left medial humeral epicondyle ㉟.

Lines: Skin creases form on the ulnar aspect of the right/left radiocarpal joint.

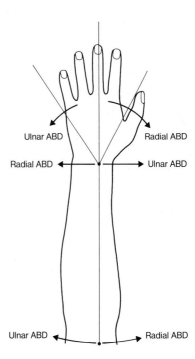

Fig. 243. Ulnar abduction/radial abduction at the joints of the hand

209

Axes: The long axes of the right/left hand (40) and the right/left forearm (36) form an ulnar angle of less than 180°.

> **Radial abduction:** 180° (between the long axes of the forearm and the hand, zero position), less the radial angle formed between the long axes of the forearm and the hand. Movement tolerance is 12°-30° (Fig. 243).

Points: In pronation dDP metacarpophalangeal joint of the right/left index finger ④ moves towards pDP right/left medial humeral epicondyle ㊱; in supination it moves towards pDP right/left radial head ㊾.

Lines: Skin creases form on the radial aspect of the right/left radiocarpal joint.

Axes: The long axes of the right/left hand (40) and the right/left forearm (36) form a radial angle of less than 180°.

5.4.5 Movements at the Phalangeal Joints

Movement Tolerances at the Joints of Fingers II- V

Metacarpophalangeal joint
Flexion/extension: 90°-0-10° to 30°.[1]
Abduction/adduction from the zero position in relation to flexion/extension. The extent of the abduction of splayed fingers is measured by the distance between the finger tips.
When the metacarpophalangeal joints are slightly flexed, the index and the little finger are able to touch and the middle finger and the ring finger can be crossed.

Proximal interphalangeal joint (PIP)
Flexion/extension: 100°-0-0°.

Distal interphalangeal joint (DIP)
Flexion/extension: 80° to 90°-0-0° to 10°.

Movement Tolerances at the Joints of the Thumb

Carpometacarpal joint (CM)
Radial abduction/adduction in the palmar plane: 70°-0-0°. Transpalmar adduction to 30°.

---

[1] In notation, the angle of flexion is noted first, then the zero position and finally the angle of extension.

210

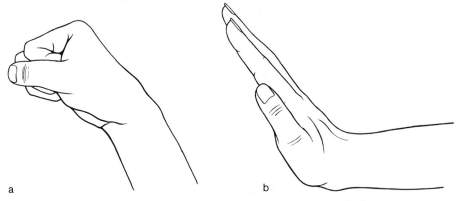

Fig. 244a, b. Clenching the fist and opening it out in a co-rotational continuing movement

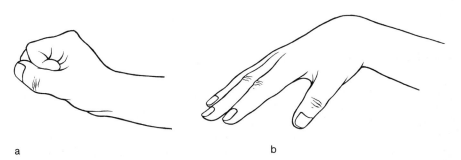

Fig. 245a, b. Functional clenching of fist and functional opening it out

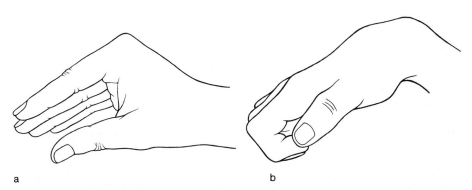

Fig. 246a, b. Fine-movement position of the hand and claw hand

Palmar abduction/adduction perpendicular to the palmar plane in a palmar direction: 70°-0-0°. Circumduction with rotation (opposition/retroposition).

Metacarpophalangeal joint (MP)
Flexion/extension: 50°-0-0° to 30°.

Interphalangeal joint (IP)
Flexion/extension: 80°-0- to 10°.

Test Position of the Hand

Clenching the fist and opening it out in a co-rotational continuing movement (as in PNF patterns) (Fig. 244).
Clenched fist: All joints of all the fingers and the wrist are flexed. Full flexion is not possible because of active insufficiency of the flexors and passive insufficiency of the extensors.
Opening the fist: All joints of all the fingers and also the wrist are extended. Full extension is not possible because of active insufficiency of the extensors and passive insufficiency of the flexors.

Functional clenched fist and functional opening it out in a mixed continuing movement (Fig. 245).
Functional clenched fist: Flexion of all the joints of all the fingers and extension of the wrist. Full force can be activated and full flexion is possible because the flexors are proximally stretched by the proximal contraction of the extensors.
Functional opening of the hand: Extension of all joints of the fingers and flexion of the wrist. Full force can be activated and full extension of the fingers is possible because the extensors are proximally stretched by the proximal contraction of the flexors.

Fine-movement position of the hand and the antagonistic *claw hand* ('crocodile hand') for developing maximal strength and mobility in flexion at the distal two joints of the fingers (Fig. 246).
Fine-movement position: Extension of the wrist, flexion with abduction or adduction at the metacarpophalangeal joints, extension at the proximal and distal interphalangeal joints of fingers II–V, opposition of the thumb through extension at the metacarpophalangeal and interphalangeal joints.
Claw hand: Flexion of the wrist, extension with abduction or adduction at the metacarpophalangeal joints, flexion in the proximal and distal interphalangeal joints of fingers II–V, retroposition of the thumb through flexion at the metacarpophalangeal and interphalangeal joints.

These six positions of the hand will effectively show up any functional deficiencies or weakness of the hand. At the same time, repeated practise of these six positions constitutes a sound exercise programme for the hands which, once mastered, can be repeated anywhere at any time and leaves no function of the hand unexercised.

6 Functional Status

Determining a patient's functional status is the overture to treatment.

Definition: Determining functional status means collecting and evaluating data concerning the patient's constitution, mobility and postural statics, as related to a hypothetical norm and the patient's present condition.
Deviations of the constitution, mobility and statics from the hypothetical norm are pathological:

- If economical motor behaviour is affected by incorrect loading of the passive structures of the motor apparatus and abnormal muscle tone
- If, regardless of the medical diagnosis, they explain the presence of pain
- If they disrupt sensory motor development in the young

Notation
The data obtained from observing and examining the patient must be written up or *notated*. If notation is carried out according to a system, with all the relevant questions ready formulated, it is only necessary to note the deviations from the hypothetical norm in the established manner and sequence. These findings must be evaluated, in that one must try to find out the logic of how they hang together.
Interpreting the data recorded means formulating the patient's functional problem; out of this follows the design of the therapy plan – form, dosage and choice of means.

Order of notation:
- Condition
- Constitution
- Mobility
- Postural statics
- Formulation of the functional problem
- Therapy plan

6.1 Condition

Definition: In determining functional status, the influence of social position and psychic and somatic states upon the patient's motor behaviour are collectively assessed under the heading 'condition'.

Social position
- Personal data
- Insurance: Is the patient insured or paying privately?
- Occupation: Training and present employment; physical and mental strain caused by work
- Hobbies, sports, artistic activities

Psychological states
- The physiotherapist is normally informed in the referral notes of any psychological disorders for which a patient is receiving treatment. To deal with these requires specialized knowledge which we shall not discuss here.
- The physiotherapist needs to listen carefully to how the patient describes his illness; this will enable her to judge whether the patient is likely to be co-operative, an aspect of great importance when planning a course of treatment. Terms like optimistic/pessimistic, euphoric/hypochondriac, talkative/taciturn, aggressive/apathetic, exaggerating/understating, contented/discontented, humorous/hypersensitive, ingratiating/complaining, restless/phlegmatic, intelligent/foolish indicate aspects according to which, without intending any value judgement, the astute physiotherapist can try to assess her patient.
- Family environment and private problems are certain to play an important role in motivating the patient to regain his health. It should be left to the patient when – perhaps not until treatment is under way – and how much he wants to reveal about these. The assessment of the patient's psyche should not be the product of questioning, but of the physiotherapist's knowledge and experience of human beings.

Somatic state
- The medical diagnosis which has led to a course of physiotherapy being prescribed.
- Assessment of the patient's condition in relation to his real age: does he appear older or younger than his chronological age?
- Assessment of the patient's body weight with regard to his state of nutrition.
- Assessment of the musculature with regard to the amount of training or exercise it is given.
- A short medical history detailing previous treatment, outcome, operations, serious accidents, etc.
- History of pain: localization, time of occurrence, possible triggers and quality of pain:
  Where is the pain?
  When did the patient first notice the pain?
  At what time in the course of the day or night does it occur?
  Is the pain connected to physical strain? a certain posture? the weather? a certain type of movement?
  What kind of pain is it: stabbing, twingeing, dull, burning, tingling, cold, numb?
  Is the pain sporadic or constant?

Note
Once pain is recognized as a life-preserving mechanism and its cause is understood, it becomes a 'road sign' on the way to therapy.

6.2 Constitution

Definition: In determining the functional status, the influence of the *length, width, depth* and *weight* of body segments upon the patient's motor behaviour is assessed under the heading 'constitution'.

The normative proportions are quite explicitly purely hypothetical and represent average values. They form a paradigm which enables us, for instance, to ascertain that a thorax is over-long, or has a reduced width, or a shoulder girdle is carrying too much weight. It is therefore unnecessary to differentiate between the sexes.

Note
Deviations of certain lengths, widths, depths and weights from hypothetical norms alter a person's motor behaviour in a predictable manner, particularly when the long axis of the body is inclined out of the horizontal (see p. 279).

The extent of the deviation is indicated as follows: $+/-$ slight deviation; $++/--$ pronounced deviation; $+++/---$ extreme deviation.

Height, weight and bone structure: Notation

- The height is measured in centimetres.
- The body weight is given in kilograms.
- The bone structure is described as: normal/light/heavy. Pronounced differences between body segments are noted. The note 'BS arms/legs slight' means that the bone structure is normal except for the arms and legs.

From the relationship of these three values we can judge whether the body weight is normal or below or above average. If it is below average, one should often indicate whether the overall condition is poor or the musculature in a poor state of training; if it is above average one should state whether it is due to over-eating or to muscular training. Overweight due to over-eating is usually found ventrally, that due to muscular training at the shoulder girdle, arms and legs.

- *Example*
'Total body length 172 cm/weight 70 kg/bone structure: BS legs slight, otherwise normal' means that the bone structure of BSs pelvis, thorax, head and arms is normal and that of BS legs is slight, and that the body weight is thus close to the upper limit.

Lengths: Notation and hypothetical norms of lengths with reference to the distribution of weight (Fig. 247)

- The greater trochanter divides the total length of the body into a lower and an upper length, each corresponding to one-half of the total length of the body, i.e. a ratio of 1:1.

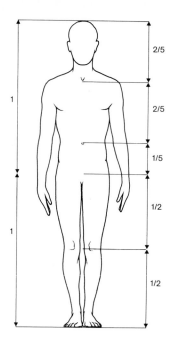

Fig. 247. Constitution: lengths

- The lower length corresponds to the length of BS legs and represents the vertical distance from the ground to the greater trochanter. The interarticular space of the knee divides the lower length in half, the distance from the ground to the interarticular space of the knee being normally a little greater than that from the interarticular space of the knee to the greater trochanter (thigh length).
- In the upright stance, the upper length corresponds to the total length of BSs pelvis, thorax and head. The length of BS pelvis, from the pubic symphysis to the navel, constitutes one-fifth of the upper length, the length of BS thorax, from the navel to the suprasternal notch, two-fifths, and the length of BS head, from the suprasternal notch to the vertex, two-fifths of the upper length of the body.
- The length of the arm, from the acromioclavicular joint to the tip of the third finger, is regarded in relation to the upper length. If the tip of the third finger reaches to about just above the middle of the thigh, the wrist joint is slightly below the greater trochanter, and this is a good norm.

For notation of deviations from the norm in lengths, we name the relevant body segment or its parts; for deviations in weight we simply name the relevant part of the body with the added designation 'weight'.

● *Examples*
'+ Upper length' means the upper length is greater than the lower length.
'Upper length (+ BS thorax)' means that within the total upper length, which is normal, the length of BS thorax is more than two-fifths.

216

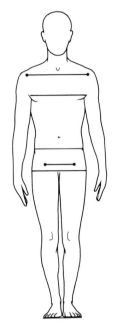

Fig. 248. Constitution: widths

'+ Upper length (+ BS thorax, + weight at the shoulder girdle)' means that the upper length is greater than the lower length, that BS thorax is more than two-fifths of the length of this over-long upper half, and that the shoulder girdle is heavier than the norm.

'– Lower length (+ thigh)' means that within the lower length, which is less than half the total length of the body, the thigh is relatively long.

Widths: Notation and hypothetical norms of widths with reference to the distribution of weight (Fig. 248).

– The track width in two-legged standing (distance between the centres of the talocrural joints) corresponds at least to the distance between the hip joints and at most to the width of the pelvis. Divergence/convergence of the long axes of the feet is assessed under the heading of statics. The width of the pelvis is taken as the distance between the two anterior iliac spines or the distance between the points where the iliac crests intersect with the midfrontal plane. It corresponds to the greatest normal track width.
– The distance between the right and left greater trochanters corresponds roughly to the frontotransverse diameter of the thorax.
– The distance between the right and left hip joints corresponds to the length of BS pelvis, or one-fifth of the total upper length, or one-tenth of the total length of the body. It is also equal to half the distance between the right and left shoulder joints.
– The distance between the right and left shoulder joints corresponds to the length of BS thorax, or two-fifths of the total upper length, or one-fifth of the total length of the body.

217

For the notation of deviations from the norm in widths, we list the above distances from the bottom upwards. When stating weights, it is a matter of judgement whether we assign them to lengths, widths or depths.

- *Examples*

'− Track width, + distance GTs' (abduction syndrome of the shoulder joints, see p. 222) means that, with a narrow track width and great distance between the greater trochanters, the arms do not hang freely.

'+ + Track width with − frontotransverse thoracic ∅' means that the patient has a narrow thorax and stands with his feet very wide apart. Whether in doing this he is compensating for a difference in leg length can only be judged when the statics have been assessed.

Depths: Notation and hypothetical norms of depths with reference to the distribution of weight (Fig. 249).

- The length of the foot corresponds approximately to the sagittotransverse diameter of the thorax.
- The ratio of the sagittotransverse diameter of thorax to the frontotransverse diameter of thorax is approximately 4:5.
- The sagittotransverse diameter of thorax is greater than the sagittotransverse diameter at the level of the navel.

Fig. 249. Constitution: depths

- The ratio of the distance between DP medial process of the calcaneal tuberosity and DP medial malleolus to the distance between DP proximal joint of the large toe and DP medial malleolus is 1:1.5 (Fig. 250a).
- The ratio of the distance between DP lateral process of the calcaneal tuberosity and DP lateral malleolus to the distance between DP proximal joint of the little toe and DP lateral malleolus is 1:2 (Fig. 250b).
- The sagittotransverse diameter of the head, from the point of the nose to the hindermost point on the head, projected into the horizontal plane, corresponds approximately to the length of the foot and the sagittotransverse diameter of the thorax. The head is normally over the feet.

For the notation of deviations from the norm in depths we list the above diameters and distances from the bottom upwards. For data about weight, we simply name the relevant part of the body with the added designation 'weight'.

● *Examples*
'+ Sagittotransverse ∅ at the level of the navel (+ + + weight in the abdomen), sagittotransverse ∅ of the head (+ + facial skeleton)' is a frequent picture in neck syndromes, when the overprominent bony structure of the facial skeleton and a comparatively small occiput cause chronic reactive hypertonia in the musculature of the neck (Fig. 251).
'− Length of foot, + sagittotransverse thoracic ∅ (+ weight in thorax)' suggests that reduced length of the foot has not left much room for weight transference forwards or backwards at the sole/ground level.

When analysing the distribution of weight, the patient's constitutional weight cannot be separated from conditional overweight. Sometimes, however, we must make it clear that we are dealing with conditional weight. The data about body size and type of bone structure provide a point of reference in regard to over- and underweight.

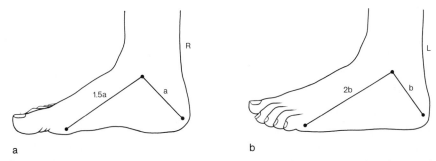

Fig. 250. Distance from the calcaneal tuberosity to **a** the medial malleolus and **b** the lateral malleolus

Fig. 251. + + Diameter at the level of the navel and + facial skeleton

Functional Consequences of Pronounced Deviations from the Hypothetical Constitutional Norm

Note

A patient's constitution *cannot be altered,* and there is no causal connection between pain and constitutional variants within the norm. Only deviations in mobility and postural statics cause pain that can be accounted for, because they disrupt economical motor behaviour and render it finally impossible (see p. 213). Pronounced constitutional deviations cause problems in everyday life in our civilized environment, which is tailored to standard sizes, forcing the individual to modify his motor behaviour to fit.

+ + +/− − − *Body size:* Adaptation to seats, working surfaces, vehicles or appliances requires lengths to be made either shorter, usually in the area of the vertebral column, or longer, which can only be achieved by using appropriate forms of padding for sitting and standing.

+ + + *Body weight:* Overweight is a conditional phenomenon, the effect of which differs from that of constitutional body weight because of where it is sited on the body segments. Excessive conditional weight is often carried ventrally on BS pelvis and BS thorax (on the abdomen and chest), but also on the buttocks and the thighs. It disturbs the body's front/back equilibrium.

+ + +/− − − *Bone structure:* Heavy bone structure means greater constitutional body weight, while lighter bone structure means less constitutional body weight. Varying bone structure in the different body segments causes changes in the proportions of weight within the body. Relatively great weight of individual body segments imposes undue strain on the segments below it. Frequent combinations are: heavy BSs pelvis, thorax and head with light BSs legs and arms (this variant means that BS legs bears a considerable load); light BSs thorax, head and arms with heavy BSs legs and pelvis (two clothing sizes bigger below the navel, a good variant for load distribution); and light BSs legs and pelvis with heavy BSs thorax, head and arms (two clothing sizes larger above the navel, a variant giving unfavourable load distribution).

+ + + *Lower length/− − −* *upper length:* Long legs combined with a short upper length are functionally advantageous for the vertebral column. When the long axis of the body is bent forwards flexionally at the hip joints, the lever arm which needs to be stabilized against gravity is short. Excessive length of the thigh within the lower length imposes extra strain upon the knee joint when bending down, unless the long axis of the body is brought as near as possible into a horizontal position flexionally at the hip joints (see p. 280).

+ + + *Upper length /− − − lower length:* Great upper length combined with short legs is functionally disadvantageous for the vertebral column. When the long axis of the body is bent forwards flexionally at the hip joints, the lever arm which needs to the stabilized against gravity is too long and needs to be shortened. In doing this, it loses its stabilization.

When *BS pelvis is overlong,* and when this is due to the length of the pelvis itself, bending the long axis of the body flexionally at the hip joints causes the lumbosacral junction to be more cranial, thus placing extra strain upon the lumbosacral area. When *BS thorax is overlong,* bending the long axis of the body flexionally at the hip joints brings into being a long lever, top-weighted by BSs arms and head, thus imposing load upon the lumbar spine. When *BS head is overlong,* the extra length tends to lie in the cervical spine, and the conditions with regard to load are relatively favourable when the long axis of the body is flexed at the hip joints.

+ + +/− − − *Arm length:* Arms are considered as long or short in relation to the total length of BSs pelvis and thorax. Where the upper arm is long or BS thorax is short, the elbow can be supported on the iliac crest with the thoracic spine in the neutral position, which can be used to relieve strain upon the paravertebral musculature. Still more important is that the total length of BSs pelvis and thorax corresponds to that of the arm from the acromion to the wrist, for then the patient can use his arms for support when sitting upright.

+ + +/− − − *Track width:* A wide track means an increase in the support area, which secures equilibrium in the frontal plane and reduces strain upon the musculature. A narrow track means a reduced support area, destabilizes the equilibrium and increases the strain upon the musculature. Too little room for the tissues on the medial aspect of the thigh hampers the potential mobility of the pelvis in the joints of the hips and the lumbar spine.

$+++/---$ *Distance between the greater trochanters:* If the greater trochanters are very far apart, there will be a varus angle of the femoral neck. When only one leg is bearing weight, the strain upon the abductor muscles at the hip joint is reduced by lengthening the lever arm and shortening the weight arm; the strain on the rotators increases. If the greater trochanters are less far apart than the norm, there will be a valgus angle of the femoral neck; the strain upon the abductor muscles at the hip joint increases, that upon the rotators decreases.

$+++/---$ *Distance between the hip joints:* If the hip joints are very far apart, the transference of weight in the frontal plane when shifting from two-legged to one-legged standing is greater than that which is necessary when the hip joints are very close.

$+++/---$ *Frontotransverse diameter of thorax:* An extra-wide thorax makes only minor weight transference necessary when changing from two-legged to one-legged standing. When the thorax is narrow, this transference can only be accomplished by lateral translation of the thorax.

$+++/---$ *Distance between the shoulder joints:* Too great a distance between the shoulder joints leads to the shoulder girdle's being poorly placed on the thorax and the pincer jaws' closing, thus increasing the demands made upon the stabilizing musculature. Too small a distance between the shoulder joints inhibits the arms from hanging freely by the side and causes abduction syndrome of the shoulder joints (Fig. 252).

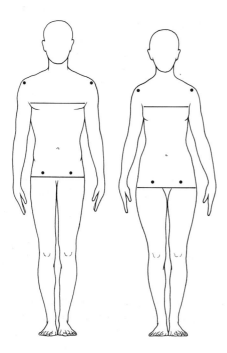

Fig. 252. Abduction syndrome of the shoulder joints

+ + +/− − − *Foot length:* A long foot increases the size of the support area forwards/backwards and the range of forward/backward transference of weight also increases. In economical gait, when the functional long axis of the foot points forwards, a long foot makes the step longer (see p. 265), and a short one makes it shorter.

+ + +/− − − *Size of the heel:* A large heel results in forward transference of the whole body weight and strain upon the forefoot. To initiate a reactive step in normal gait requires forward transference of only a little weight. A small heel results in backward transference of the whole of body weight, which is unfavourable because when strain is imposed on the heel there is no spare movement tolerance backwards unless a step backwards is taken, and considerable forward transference of weight is required to initiate a reactive step forwards.

+ + +/− − − *Sagittotransverse diameter of thorax:* An extra-deep thorax measurement is an indication of a round back or hyperkypholordosis at the thoracic level, or, more rarely, a sign that the thorax is fixed in the inspiration position. A very shallow thorax measurement indicates a flat back or a funnel chest (see p. 158).

+ + +/− − − *Sagittotransverse diameter of head:* The significance of the sagittotransverse diameter of the head is that the balance of weights ventral and dorsal of the midfrontal plane is of decisive importance to the potential mobility of the cervical spine and atlanto-occipital and atlanto-axial joints. Predominance of dorsal weight has never been observed. An overemphasized facial skeleton accompanied by a small posterior cranium, however, is a frequent constitutional deviation; attention should be paid to it in patients who suffer from headaches and cervical kyphosis.

The role of distribution of the constitutional + + +/− − − *weights:* Greatly increased or reduced weights cause an increase or decrease in the weights of levers. The weight of a lever in relation to its length is not very important when the long axis of the body is vertical, but the closer the stabilized long axis of the body is to the horizontal, the greater the effect of the weight. The increase in weight affects a distally loaded lever more than a proximally loaded lever. A heavy shoulder girdle imposes more strain on the vertebral column (top-heaviness) than a heavy pelvis.

Weight in widths and depths: Asymmetrical distribution of weight across widths (i. e. in relation to the symmetrical plane) and along depths (i. e. in relation to the midfrontal plane) has an unbalancing effect upon the passive structures of the vertebral column when the long axis of the body is vertical. These are problems of postural statics. The body reacts to weight imbalance with hypertonia of the musculature on the opposite side, the normal equilibrium reaction of a healthy body to poor posture. The other reaction of the body is buttressing with a counterweight, which subjects the passive structures of the motor apparatus to shearing stress. Both these reactions can trigger pain, which thus has its origin in the postural statics, but these are in turn inseparable from the constitution.

6.3 Mobility

Definition: In determining the functional status, the influence of the passive and active movement tolerances at the joints upon the patient's postural statics and motor behavior, with reference to constitution and condition, are assessed under the heading 'mobility'.

To reach end-stop when testing a joint's mobility, the braking activity of muscles must be cut out, either by reducing weight and/or by making allowance for the passive insufficiency of multi-joint muscles.

We use the points, lines and axes listed in the chapter on functional measuring. The normative values for movement tolerances are clearly set out by Debrunner (1971). We use X to notate restriction of movement (X, slight restriction; XX, considerable restriction; XXX, total restriction) and 8 to express hypermobility (8, slight hypermobility; 88, considerable hypermobility; 888, extreme hypermobility). Following the AO system of joint measurement, we express movement tolerances at joints of the extremities in degrees.

Criteria for the mobility of the vertebral column
- Can BS pelvis, BS thorax and BS head be easily aligned in the long axis of the body?
- If yes, can they be kept there?
- If not, in what motion segment or at what level of movement does the restriction or instability lie?

Criteria for the mobility of BS legs
- Can the hip joints reach the neutral position? What is the tolerance for extension? Can the long axis of the thigh be first vertically aligned and then inclined slightly forward in standing?
- Can the knee joints reach the neutral position? Does the patient suffer from hyperextension of the knee? What is the tolerance for flexion?
- Can the talocrural joints reach the neutral position? What is the tolerance for dorsiflexion? Can the long axis of the lower leg be first vertically aligned and then inclined slightly forward, keeping the soles of the feet in contact with the ground?
- Are the subtalar, tarsometatarsal and phalangeal joints mobile enough to form the normal longitudinal arch of the foot, if necessary by manipulation?

Criteria for the mobility of BS arms
- Can all the joints reach the neutral position? Is the movement tolerance at all joints sufficient for the arms to assume their role as reactive pendula in normal gait?

Optimal erect posture and economical employment of the fall-preventing musculature are only possible if movement tolerances are as described in Chap. 5.

Examination and Notation of the Mobility of the Vertebral Column

Flexion (FLEX)

Starting position 1: Sitting upright on a stool (Fig. 253). The first stage tests the lumbar and thoracic spine. The patient is asked to let his body 'collapse'. This causes extension of the pelvis at the hip joints, flexion in the lumbar and thoracic spine, and extension in the cervical spine. In this posture, the vertebrae can be counted by their spines, and it is easy to observe the localization of restrictions of movement and hypermobility. In the second stage the patient is asked to move DP point of chin towards DP suprasternal notch, so that the flexion of the cervical spine can also be assessed. The weight of the head shifting forwards causes the pelvis to move backwards extensionally at the hip joints and the lumbar and thoracic spine to go into further flexion (buttressing by counterweighting). If the patient is not able to cut out the braking activity of the extensors, the physiotherapist takes over the weight, gripping the thorax firmly on both sides (Fig. 254).

Starting position 2: On all fours across the treatment bench (Fig. 255). Because of the bridging activity of the abdominal musculature, the pelvis is extensionally activated at the hip joints and flexionally activated at the lumbar and thoracic spine. For assess-

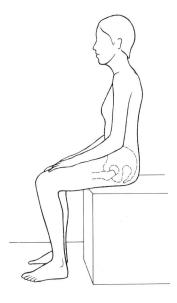

Fig. 253. Sitting upright on a stool (starting position 1)

225

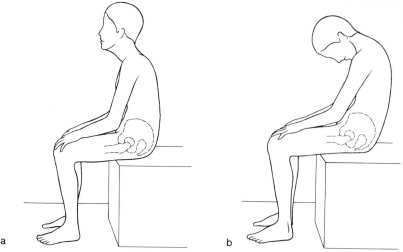

Fig. 254. a Phase 1: Flexion of lumbar and thoracic spine, extension of pelvis at hip joints.
b Phase 2: + Flexion of cervical spine from upright sitting

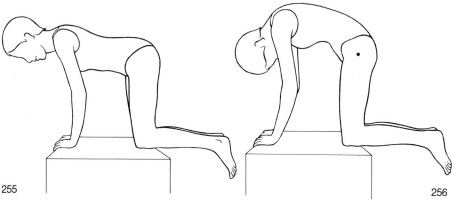

255

256

Fig. 255. On all fours across the treatment bench (starting position 2)

Fig. 256. Flexion of lumbar, thoracic and cervical spine from the all-fours position

ment of the cervical spine, additional flexion of that region can be initiated from DP vertex (Fig. 256).

Extension (EXT)

Starting position 1: Sitting upright on a stool (Fig. 257a). The first stage tests the lumbar and thoracic spine. The patient inclines the stabilized long axis of his body forwards at the hip joints to about 40° and places his hands on his thighs at a place below the shoulder joints, with the fingers facing medially and BS arms in supporting function.

226

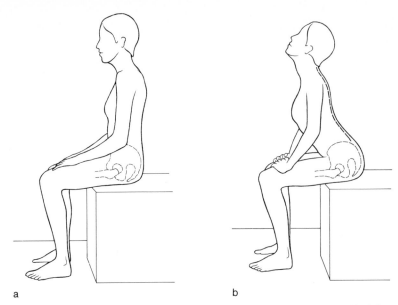

Fig. 257a,b. Extension of lumbar, thoracic and cervical spine and flexion of pelvis at hip joints from upright sitting

He is asked to 'make creases in the skin' at the lumbosacral junction. As he does so, the pelvis is flexed at the hip joints, with concentric isotonic extension of the lumbar and thoracic spine. Activity and movements can now be easily assessed. In the second phase the patient is asked to 'make the front of his neck longer, open his mouth and look at the ceiling'. We look particularly closely at the lower part of the cervical spine, which frequently shows a neck kyphosis and therefore does not participate in the extension at all (Fig. 257b).

Starting position 2: On all fours across the treatment bench (Fig. 258a). The patient is asked to 'drop his back' and activate total extension of the vertebral column by raising his head and looking forwards and upwards. In doing, this, the pelvis moves flexionally at the hip joints. The physiotherapist can now test particular motion segments of the vertebral column, manipulating the weight of BSs pelvis and thorax from below, inducing flexion through raising and extension through lowering (Fig. 258b).

Notation
'FLEX LS, CS X/EXT TS 8' means: The flexion of the whole of the lumbar and cervical spine is slightly limited; the whole of the thoracic spine has slight hypermobility in extension.
'FLEX LS lower X, TS (kyphosis caudally extended to L2)/EXT T10–L2 X' means: The lower part of the lumbar spine is slightly limited in flexion; the two upper segments, by contrast, show abnormal kyphosis. Segments T12–L2 have slightly limited extension.

227

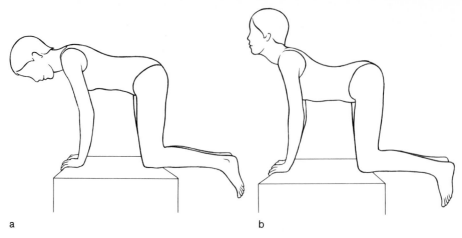

a   b

Fig. 258 a, b. Extension of lumbar, thoracic and cervical spine and flexion of pelvis hip joints from the all-fours position

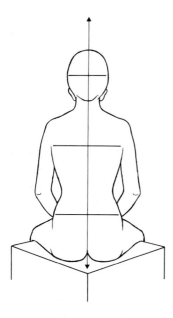

Fig. 259. Sitting upright on a stool (starting position 1)

Lateral flexion (LATFLEX)

When we speak of left lateral flexion, we mean left concave. When we speak of right lateral flexion, we mean right concave. In a lateroflexional movement the distance points are on the concave aspect, the vertex of the curvature on the convex aspect.

Starting position 1: Sitting upright on a stool (Fig. 259), legs comfortably apart. To test *right concave lateral flexion of the lumbar spine* the patient – particularly if his thorax is

228

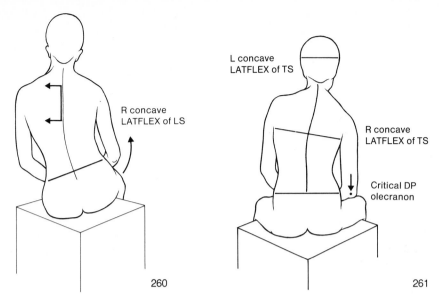

260

261

Fig. 260. Right concave lateral flexion of the lumbar spine as the thorax moves left in translation

Fig. 261. Right concave lateral flexion of the thoracic spine; left concave lateral flexion of the cervical spine constitutes a counter-rotational continuing movement of DP olecranon downwards and slightly to the right

of small frontotransverse diameter – should move his thorax so far to the left that DP right ischial tuberosity loses contact with the seat. As he does this, there is inward rotation of the pelvis at the left hip joint and right concave lateral flexion of the lumbar spine (Fig. 260). The opposite happens when the thorax is moved to the right. If the patient cannot keep the frontotransverse diameter of the thorax horizontal, the physiotherapist assumes the weight by gripping the thorax firmly on each side, manipulates the necessary gliding movement of the thorax to the left and supports the raising of the pelvis from its base support. For *right concave lateral flexion of the thoracic spine* the patient moves DP olecranon downwards and a little to the right laterally (Fig. 261). The pressure exerted on the seat does not change; left concave lateral flexion arises in the cervical spine. If the cervical spine is stiff, DP vertex is also directed to move downwards to the right laterally (and vice versa for the other side). In testing *right concave lateral flexion of the cervical spine* the patient should support his head by placing his right hand under his right ear while DP earlobe moves closer to DP right acromion and DP vertex moves downwards and to the right laterally (Fig. 262; vice versa for the opposite side). It is important that the frontotransverse diameter of the thorax maintains its horizontal position. Figure 263 illustrates a variant with co-rotational continuing movement of the left arm in the mid frontal plane.

Starting position 2: On all fours (Fig. 264a). To test left concave lateral flexion of the lumbar and thoracic spine together, DPs right and left heel are moved laterally/to the left/cranially. The continuing movement brings about left concave lateral flexion of the lumbar spine. As a buttressing movement, DP left acromion moves laterally/to the

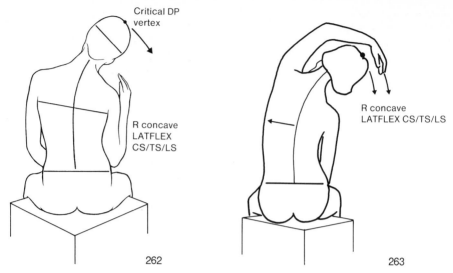

Fig. 262. Right concave lateral flexion of the cervical, thoracic and lumbar spine, constituting co-rotational movement starting from DP vertex, out of the upright sitting posture

Fig. 263. As Fig. 262, with co-rotational movement of the left arm

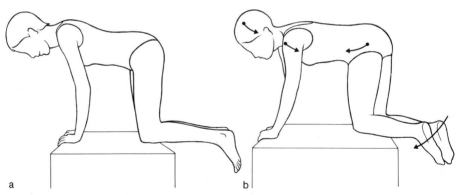

Fig. 264. a Starting position on all fours. **b** Left concave lateral flexion of lumbar, thoracic and cervical spine; buttressing through counter-movement

left/caudally and brings about left concave lateral flexion of the thoracic spine. To test the cervical spine, DP vertex moves laterally/to the left/caudally. DP T10 moves to the right, with left concave lateral flexion in the lumbar and thoracic spine. All these are repeated for the other side. An examination of this kind shows restrictions of latero-flexional movement in scoliotic posture with great clarity (Fig. 264b).

Notation
'LATFLEX LS R concave, TS L concave, CS R concave X' means: Right concave lateral flexion in the lumbar spine, left concave lateral flexion in the thoracic spine and right concave lateral flexion in the cervical spine are slightly restricted.

230

'LATFLEX L concave L2–T9 88, avoidance mechanism translation' means: Left concave lateral flexion in the area L2–T9 is markedly hypermobile and shows an avoidance mechanism in the form of a translation of the thorax (gliding body) to the right.
'LATFLEX R concave S1–L4 XXX, L concave T3–C5 XXX' means: In right concave lateral flexion the motion segments S1–L4 are totally restricted; in left concave flexion the motion segments T3–C5 are totally restricted.

Positive/negative rotation (+ / − ROT)

The levels of rotation in the spine lie between the pelvis and the thorax and between the thorax and the head. By 'rotation level thoracic spine' we mean 'lower thoracic spine/upper lumbar spine', and by 'rotation level cervical spine' we mean 'cervical spine and atlanto-occipital and atlanto-axial joints'. In addition to specifying the level of rotation, the rotary pointer and the direction in which the pointer moves should also be specified.

1. Rotary pointer pelvis or line connecting iliac spines
2. Rotary pointer thorax or frontotransverse diameter of thorax
3. Rotary pointer head or line connecting earlobes

● *Examples*
'ROT TS thorax positive' means that the pelvis remains stationary while the thorax performs a positive rotation at the level of the thoracic spine.
'ROT TS pelvis negative' means that the thorax remains stationary while the pelvis performs a negative rotation.
'ROT TS and CS, thorax negative' means that the pelvis and head remain stationary while the thorax performs negative rotation at the level of the thoracic and cervical spine. The thorax is the cranial pointer for the thoracic spine and the caudal pointer for the cervical spine.

Starting position: Sitting upright on a stool (Fig. 265). To test the rotation, BSs pelvis, thorax and head must be aligned in the vertical long axis of the body. For economical activity to be possible, there must be potential mobility for flexion and extension of the pelvis at the joints of the hips and lumbar spine, dynamic stabilization of the thoracic spine in the neutral position, and potential mobility of the head at the atlanto-occipital and atlanto-axial joints and the joints of the cervical spine. The vertebral column can only rotate freely and lift-free when the tone of the superficial musculature of the abdomen, back and shoulder girdle has been considerably reduced (see p. 322).

Testing rotation in the thoracic spine: The frontotransverse diameter of the thorax is the cranial rotary pointer for rotation level thoracic spine and the caudal pointer for rotation level cervical spine. As the movement tolerance of the cervical spine and the atlanto-occipital and atlanto-axial joints is about twice that of the thoracic spine, the head can remain in one position during testing of the thoracic spine. The crossed hands rest on the sternum so as not to inhibit the rotation. The patient is given the following instruction: 'Look straight ahead. This movement should not change the pressure exerted on the seat in any way. Your hands do not alter their position on your chest but turn

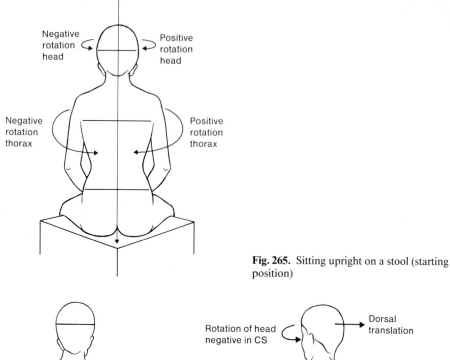

Fig. 265. Sitting upright on a stool (starting position)

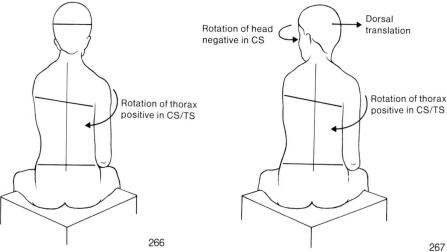

266

267

Fig. 266. The thorax has performed positive rotation in the thoracic and cervical spine. The head is a fixed point in space

Fig. 267. The thorax has performed positive rotation in the thoracic and cervical spine; negative rotation of the head in the cervical spine

with the thorax alternately to the right and left' (Fig. 266). The physiotherapist observes the movement of the frontotransverse diameter of the thorax in the horizontal transverse plane in relation to the line connecting the iliac spines, which is stationary. When, as illustrated in Fig. 266, the full tolerance of rotation in the thoracic spine has been exhausted, there is plenty left in the cervical spine and the atlanto-occipital and atlanto-axial joints. This can be exhausted to end-stop by additional negative rotation of the head in the cervical spine, a buttressing countermovement which thus actively buttresses the rotation in the thoracic spine (Fig. 267).

Avoidance mechanisms frequently appear in positive rotation of the thorax in the thoracic spine:

1. Translation of the thorax to the left, although this is accompanied by an increase in the pressure of the left ischial tuberosity on the seat
2. Retraction of the right scapula, often combined with translation of the thorax to the left
3. Right concave lateral flexion in the thoracic spine.

Since well-functioning rotation in the thoracic spine largely depends upon posture, the physiotherapist should manipulate the movement if the test result is poor. Standing in front of the patient, she should raise the patient's hands and thoracic cage slightly and at the same time support the upper lumbar and lower thoracic spine dorsally.

Testing rotation of the cervical spine and the atlanto-occipital and atlanto-axial joints:
The instruction is as follows: 'Look backwards over your left shoulder. The pressure exerted by your body on the seat does not change; your thorax and hands turn in the opposite direction, to the right' (Fig. 267).
Frequently occurring *avoidance mechanisms* are:

1. Ventral translation of the head
2. Left concave lateral flexion of the cervical spine in negative rotation of the head. If the physiotherapist is manipulating this rotation, she stands in front, holds the patient's head with one hand on the zygomatic arches, and during the rotation effects a dorsal translation and lateral flexion with the concavity on the opposite side to the direction of rotation, i. e., here, on the right.

Notation
'ROT TS thorax positive X, AM translation negative' means: The positive rotation of the thorax in the thoracic spine is slightly restricted; during a negative rotation the thorax effects translation to the right in an avoidance mechanism.
'ROT CS head negative, CS XXX' means: When the head is rotated to the left, rotation occurs only at the atlanto-occipital and atlanto-axial joints, not in the cervial spine.
'ROT TS thorax positive XX, AM retraction; thorax negative level of rotation shifted cranially' means: In rotation of the thorax to the right, the rotation is considerably restricted; in an avoidance mechanism the right shoulder-blade is retracted. In rotation to the left, the level of rotation is shifted cranially.

Translations

The functionally important levels of translation of the vertebral column are in the upper lumbar, lower thoracic and cervical vertebral regions. Movement excursions at these levels (see p. 163) are possible in practically all transverse directions, which we schematize as: right/left, laterally/medially, and ventrally/dorsally. We classify them as avoidance mechanisms when they occur instead of flexion, extension, lateral flexion or rotation, or when they are involuntarily mixed with these qualities of movement.

In the upright posture, translations of BSs thorax and head bring about the most unambiguous transference of weight in the horizontal plane. Flexion and lateral flexion always have one downward component and one in the opposite direction, while in rotations weight on the opposite side is moved in the opposite direction.

Notation of translations

A hypermobile, unstable vertebral column is recorded as, e. g., 'Translation LS, TS 88' or 'Translation painful'.

Examination and Notation of the Mobility of the Hip Joints

Because of the topographically dictated dependence of the movements of one hip on those of the other and of the lumbar spine, we examine these joints in relation to each other, in terms of a continuing movement and its active buttressing. The examination is performed by the physiotherapist, who uses manipulation and carries some of the weight. The patient is told about the movement of distance points and the positional changes of axes, so as to cut out the braking activity of his muscles.

Flexion/extension (FLEX/EXT)

Starting position 1: Lying supine; BSs pelvis, thorax and head are aligned horizontally in the long axis of the body. The line connecting the right and left iliac spines and the frontotransverse dimeter of the thorax are at right angles to the long axis of the body and remain parallel during the entire examination. Both hip joints are in the neutral position. The examination includes flexion, adduction, external rotation of the femur at the right hip joint and extension of the pelvis at the left hip joint. The physiotherapist assumes the weight of the right leg and guides the right knee, as the distal distance point of the right hip joint, in the direction of the left shoulder joint in the plane of a body diagonal (right hip/left shoulder joint). The right knee joint is flexed to 90° and is neutral as to rotation, which cuts out the ischiocrural brake. The long axis of the lower leg is an ideal rotary pointer for the right hip joint. As soon as the continuing movement has started, it is important to watch that the pelvis moves only extensionally at the left hip joint. This causes the curvature of the lumbar spine to alter: the line connecting the right and left iliac spines performs a parallel translation in relation to the frontotransverse diameter of the thorax and moves cranially/dorsally. To help the movement flow smoothly, the physiotherapist brings about sufficient adduction of the femur at the right hip joint to prevent the line connecting the right and left iliac spines from rotating in its transverse plane, and sufficient external rotation to prevent it from rotating or gliding in its frontal plane. The left leg must not alter its position.

With a movement tolerance at the hip joints of 120°–130° flexion and 10°–15° extension, we should ask the following questions when examining the patient:

- Can the patient take up the starting position, with all body segments parked (see p. 85)?
- Is the starting position hindered by a flexion contracture at one or both hip joints?
- Is the starting position hindered by restricted flexion in the lumbar spine or restricted extension in the thoracic spine?

If the patient is hampered in either of the latter two ways, the starting position should be modified by the use of the pillow to support the leg or legs or the lumbar spine. When evaluating the movements, these modifications must be taken into account.

With optimum flexion, adduction and external rotation at the right hip joint and elasticity of the dorsal musculature, and with optimum extension at the left hip joint and sound flexional mobility of the lumbar spine, the right knee can easily be pressed upon the thorax while the left leg remains where it was in the starting position (Fig. 268).

Flexion contracture at the left hip joint is shown by the left leg's rising as soon as the pelvis begins to participate in the continuing movement (Fig. 269).

Extension contracture at the right hip joint is shown by premature onset of the continuing movement, in which the pelvis participates (Fig. 270).

Extension contracture in the lumbar spine is shown by the rising of the whole BS pelvis, together with the left leg, as soon as the pelvis begins to participate in the continuing movement (Fig. 271).

Starting position 2: Sitting upright on a stool. To test pure flexion, the long axes of the thighs are set parallel and approximately sagittotransverse. The functional long axes of the feet (see p. 265) thus point forwards. The proximal lever increases the flexion beyond the existing 90°. BSs pelvis, thorax and head, stabilized and aligned in the long axis of the body, flex en bloc from the hip joints as far as possible without flexional deformation of the lumbar spine. The arms may be placed on the thighs to carry the weight of the thorax and head. With extensive transverse abduction of the thighs at the hip joints and no rotation, the flexion can be increased.

Notation
'FLEX/EXT R 100-0-0, L 120-0-5' means that the right hip joint has 20° less flexion than the left and no extension, but does just reach the neutral position. These ratings allow normal function of the left hip joint.

'FLEX/EXT R 90-10-0' means that flexion of the right hip joint reaches just 90° with a flexion contracture of 10°. These ratings make normal upright standing impossible.

Abduction/adduction (ABD/ADD)

The Neutral-0-Method defines the neutral position of the hip joints in regard to abduction/adduction as the position at which a right angle is formed between the line connecting the right and left anterior superior iliac spines and the line connecting one iliac spine and the lateral femoral epicondyle of the same side. As the constitutional variants of the distance between the right and left iliac spines are very great, we define the neutral position of the hip joint in regard to abduction/adduction as the position in which a right angle is formed between the line connecting the right and left iliac spine and the functional long axis of the thigh (the centre of the head of the hip joint/centre of the patella), which is the same as the rotation axis of the hip joint.

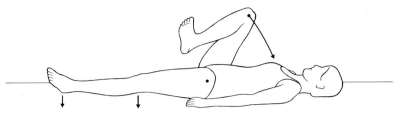

Fig. 268. Optimal flexion, adduction and external rotation of the leg at the right hip joint, extension of the pelvis at the left hip joint

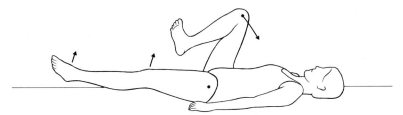

Fig. 269. Flexion contracture at the left hip joint

Fig. 270. Extension contracture at the right hip joint

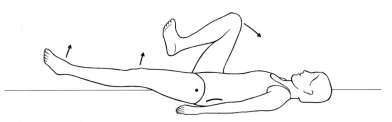

Fig. 271. Extension contracture in the lumbar spine

236

Movement tolerances from the neutral position: 30°–50° abduction, 20°–30° adduction.

Starting position 1 to test abduction in the right hip joint: Lying supine. DP right heel is guided laterally/to the right/cranially, causing abduction at the right hip joint effected by the distal lever, the leg. This causes movement of the pelvis in the form of a co-rotational continuing movement. The right iliac spine moves cranially/medially, causing right concave lateral flexion in the lumbar spine. The left iliac spine moves caudally/laterally in relation to the left hip joint and, as the proximal lever, effects abduction of the pelvis at the left hip joint. The physiotherapist performs this movement sequence by manipulation, assuming the entire weight of the patient's right leg and telling the patient where the distance points are moving. At the same time, the patient should palpate both his iliac spines. Giving this information to the patient engages his perception and disengages the braking activity of his abductors (Fig. 272).

Starting position 2 to test adduction at the right hip joint: The left leg is crossed over the right so that the left foot is lateral to the right knee. If that is not possible, the left leg, flexed at the hip and knee joints, should be raised from the base support and held by the physiotherapist. DP right heel is guided medially/cranially; this causes adduction at the right hip joint effected by the distal lever. The pelvis is set in motion in a co-rotational continuing movement. The right iliac spine moves caudally/medially, causing left concave lateral flexion in the lumbar spine. The left iliac spine moves cranially/medially in relation to the left hip joint, causing adduction/external rotation

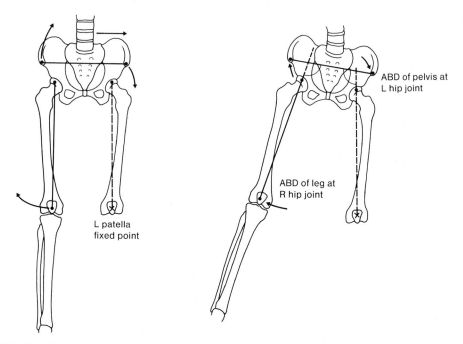

Fig. 272. Abduction at the right hip joint effected by the distal lever and abduction at the left hip joint effected by the proximal lever; right concave lateral flexion of the lumbar spine

237

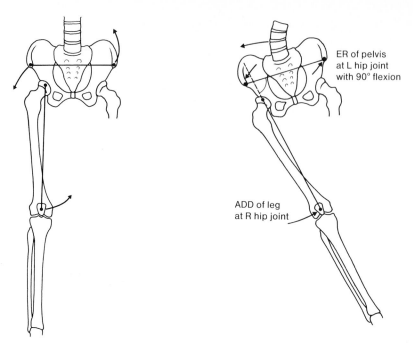

Fig. 273. Adduction at the right hip joint effected by the distal lever and external rotation at the left hip joint effected by the proximal lever; left concave lateral flexion of the lumbar spine

at the left hip joint, depending upon the extent of flexion (Fig. 273). For this test the leg can simply be pushed across the base support; again the patient needs to be told what is happening so that the braking activity of the abductors and internal rotators at the hip joints is disengaged.

Notation
'ABD/ADD R 30-0-10' means that the patient has normal but slight abduction and restricted adduction at the right hip joint.
'ABD/ADD R 50-0-30, L 0-5-30' means that the patient has normal abduction and adduction with wide movement tolerance at the right hip joint, while the left hip joint shows an adduction contracture of 5°, so that, although he has a good range of adduction, he cannot not achieve the neutral position.

Transverse abduction/adduction

Transverse abduction/adduction is measured from an angle of 90° flexion at the hip joints with zero rotation.

Movement tolerances: Transverse abduction 60°, transverse adduction 30°.

Starting position 1: Sitting upright on a stool of a height to allow the long axes of the thighs to be sagittotransverse; there should be contact between the soles of the feet and

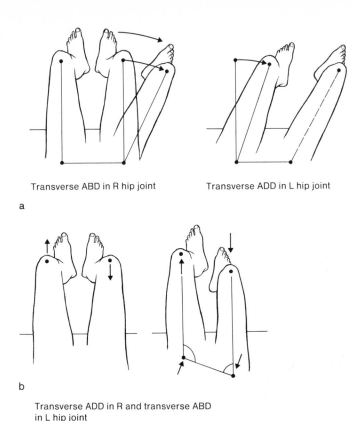

Transverse ABD in R hip joint   Transverse ADD in L hip joint

a

b

Transverse ADD in R and transverse ABD
in L hip joint

Fig. 274a, b. Starting position sitting upright on a stool. **a** Transverse abduction at the right hip joint effected by the distal lever and transverse adduction at the left hip joint effected by the distal lever. **b** Transverse adduction at the right hip joint and abduction at the left hip joint effected by displacement of the fulcrum

the floor. DP right knee moves laterally/to the right/dorsally while the sole of the right foot glides across the floor in a circular movement in the same direction. This results in transverse abduction at the right hip joint effected by the distal lever. The pelvis should not be involved by a continuing movement. Now DP left knee moves medially/dorsally, the sole of the left foot gliding across the floor in a circular movement in the same direction. The result is transverse adduction at the left hip joint effected by the distal lever (Fig. 274 a).

Starting position 2: As for starting position 1. DP right iliac spine moves dorsally/laterally while the right ischial tuberosity glides dorsally/medially across the seat in a circular movement. The right knee and talocrural joints move slightly extensionally and plantar-flexionally. Simultaneously, DP left iliac spine moves ventrally/medially, while DP left ischial tuberosity glides ventrally/medially across the seat in a circular movement. The left knee and talocrural joint move slightly flexionally and dorsal-extensionally. This brings about a positive rotation of the pelvis in the thoracic spine. Transverse abduction by displacement of the fulcrum has taken place at the right hip joint and transverse adduction by displacement of the fulcrum has taken place at the left hip joint, making a buttressing movement involving three moving points. The long axes of the thighs remain parallel (Fig. 274 b).

Notation
'Transverse ABB/ABD R 15-0-30' means that although there is considerable restriction of transverse abduction at the right hip joint, transverse adduction can be performed freely.
'Transverse ABB/ABD L 0-5-15' means that the left hip joint has a transverse adduction contracture of 5° and that transverse adduction is also restricted.

Internal rotation/external rotation (IR/ER)

Rotation of the leg and/or the pelvis about the functional long axis of the thigh is an important and economical kind of movement at the hip joints. The great tolerance for extensive, unhampered rotation permits constant change of the areas under stress, as the femoral heads glide within the hip joints, and of the muscle activities of moving, lifting, braking and prevention of falling.

Movement tolerances: With the hip joint in the neutral position, internal rotation is 40°–50°, external rotation 30°–40°. That the internal rotation is 10° greater is due to the antetorsion at the proximal end of the femur, which is normally 12° (see below). Reduced detorsion only apparently increases the internal rotation of the hip joint and reduces the external rotation. In functional observation we must note that, in the zero position, the flexion/extension axis of the knee joint is frontotransverse – but this is a functionally adjusted zero position, in which the hip joint is as far externally rotated in relation to the midfrontal plane as it normally, in its own neutral position, is internally rotated. With 90° flexion of the hip joint internal rotation is 30°–45° while external rotation amounts to 40°–50°. With the patient seated, we can observe by palpation that the distance between the greater trochanter and the seat can vary. When the lower leg is not vertical but hangs internal-rotationally from the pelvis, with the heel lateral in relation to the knee, we can see that this distance is quite wide. In such cases there is little transverse abduction, so in flexion at the hip joints effected by the proximal lever the thighs are in the way of the pelvis, and instead of flexion at the hip joints there is flexion in the lumbar spine. When the greater trochanter is not far off the seat and there is wide transverse abduction, the pelvis can be brought between the thighs during flexion at the hip joints effected by the proximal lever, and the lumbar spine can remain stabilized in the neutral position.

Starting position 1: Lying prone on a treatment bench. If the patient has a flexion contracture in one or both hip joints, BSs pelvis and thorax must be suitably supported, while the head can be parked with the forehead resting on the forearms. The physiotherapist performs the examination by manipulation, telling the patient where the distance points are to be moved. The patient may be able to palpate the iliac spines himself. The line connecting the right and left iliac spines is the proximal rotary pointer, the long axis of the lower leg the distal rotary pointer. The patient's feet should hang out over the end of the bench so that the functionally neutral position of the hip joints can be maintained. The right knee joint should be flexed to 90°, moving the long axis of the lower leg as distal pointer to lie sagittotransverse. The functional long axis of the foot (lateral aspect of heel/proximal joint of large toe) is arranged frontosagittally. Excessive divergence of the functional long axis of the foot may indicate external rotation at the knee joint or abnormally pronounced tibial torsion (see p. 271), while convergence indicates less than the normal tibial torsion. If the sole of the foot slants from lateral/cranial towards medial/caudal, there is insufficient pronation of the forefoot, characteristic of pes adductus.

DP right heel is guided laterally/to the right/downwards. The iliac spines do not move. The right greater trochanter moves ventrally/medially in relation to the right iliac spine, while the right leg has performed an internal rotation at the right hip joint. To make the movement as purely rotational as possible, the 90° angle between the thigh and the lower leg must be maintained. Because the thigh is conical in shape, the path it traces as it rolls across the treatment bench deviates slightly caudally out of a simple right/left direction.

Starting position 2: Sitting on a high stool, so that the knees are slightly lower than the hip joints. To test internal rotation on the right, the patient supports his right knee with his left; the physiotherapist manipulates the internal rotation, guiding DP right heel laterally/to the right/upwards. To test external rotation the right leg is crossed over the left and the long axis of the lower leg is arranged as near horizontal as possible without allowing the right iliac spine to move laterally/to the right/downwards.

Note
Asymmetrical rotation at the hip joints often disturbs the economy of the statics considerably. A difference between the angles of the femoral necks is often the cause of a discreet drop of the pelvis on one side.

Notation
'IR/ER R 20-0-50, L 55-0-15' means that the range of rotation is normal at both hip joints but also that the right hip joint is an example of the external rotation type and the left an example of the internal rotation type. If the pelvis is raised on the left, incomplete detorsion and incomplete detorsion of the right femoral neck in relation to the hip joint may be assumed. During the stance phase in walking the right knee joint tends to point more laterally and the left more medially (see *Gangschulung*).

If it is important to examine the joints of other extremities during mobility testing, reference can be made to the distance points, levers, pointers and pivots listed in Chap. 5.

Examination and Notation of Lumbar Kyphosis and Ischiocrural Shortening

We differentiate between structural and functional lumbar kyphosis.
Structural lumbar kyphosis is caused by the absence of lumbar lordosis and kyphotic deformation of the lumbar spine, particularly from L5 to the lower thoracic spine, usually due to rickets. In this type of kyphosis, even when the ischiocrural musculature

is relaxed – as, for instance, in sitting – the lumbar spine cannot be approximated to the normal shape by extensional movement.

Functional lumbar kyphosis is often found to varying degrees in totally flat-backed patients (see p. 158) in conjunction with ischiocrural shortening and marked antetorsion at the hip joints. When the ischiocrural musculature is relaxed this kyphosis can be partly or totally eliminated. Typically, there is lordotic deformation of the lower part of the thoracic spine, which adjoins the lumbar kyphosis cranially.

Lumbar kyphoses are examined from three different starting positions.

Starting position 1: Long sitting with the legs stretched out, knee joints at neutral, legs flexed to 90° and comfortably abducted at the hip joints. The arms are supported on the floor. The points of contact between the floor and the palms of the hand or the dorsal aspects of the proximal phalanges II–V, with the thumb pointing forwards, are by or slightly to the rear of the right and left greater trochanters. The extension of the lumbar spine is carried out through bridging activity and any kyphosis can be localized or suspected lateroflexional and/or rotational deviation ascertained.

Starting position 2: Stride sitting on a treatment bench, legs spread wide, with extensive transverse abduction at the hip joints. The lower legs hang freely over the edge of the bench. The ischiocrural musculature is relaxed. The arms are as in starting position 1. Again, the lumbar spine is moved extensionally by bridging activity, and we can ascertain whether and how far the kyphosis has been eliminated.

Starting position 3: Sitting cross-legged on the treatment bench, arms as in starting positions 1 and 2. The ischiocrural musculature is relaxed. The hip joints are in extensive transverse abduction and maximum external rotation. We ascertain whether the lumbar kyphosis can be eliminated by extension of the lumbar spine. If the angle of antetorsion is too great and the neck of the femur shows valgus deviation, sitting cross-legged is uncomfortable and not fully possible.

Notation

'Lumbar kyphosis: + + extending to T11. Cross-legged sitting: +/stride sitting: normal' means that in long sitting there is a pronounced functional lumbar kyphosis in which ischiocrural shortening plays a role; we can tell this because the kyphosis is eliminated in stride sitting. Because it is also present in cross-legged sitting, although to a lesser degree, we assume this is an internal rotation type with insufficient detorsion and shortening of those muscles at the hip which bridge only one joint. This assumption should be confirmed by testing rotation at the hip joints.

'Lumbar kyphosis: + + R concave lateroflexional, extending to L1. Cross-legged sitting: + + +/stride sitting: + +' means that this patient has pronounced structural kyphosis of the entire lumbar spine. This kyphosis is more pronounced in cross-legged sitting because there is additionally considerable antetorsion, which is not eliminated by lumbar extension in stride sitting.

Examination and Notation of Ischiocrural Shortening and Functional Stretching

Starting position: Lying supine on a treatment bench. The patient clasps his hands around his right thigh and pulls it as close as possible to his trunk. The knee joint goes into maximum flexion (Fig. 275).

Fig. 275. Functional stretching of the ischiocrural musculature

Movement sequence for ischiocrural stretching: Three movement excursions should occur simultaneously: (1) DP right heel moves ventrally/cranially, (2) DP right knee moves ventrally/caudally, and (3) DP linked hands moves caudally. Spatial fixed point: DP C7; *conditio:* the pressure exerted by C7 upon the base support must remain constant.

Analysis: With moderate extension at the right hip joint and full extension at the knee joint, effected by the distal lever, the long axis of the right leg has moved through space into the vertical – in relation to the body, into a position at right angles to the midfrontal plane. During this movement, normal stretching of the ischiocrural muscles has taken place. The activities were in (1) the quadriceps, which used the stretched ischiocrural musculature as resistance, and (2) the extensors of the thoracic spine, which, providing active buttressing, stabilized the thoracic spine in neutral against the resistance of the stretched ischiocrural muscles. This movement sequence is performed with BS arms in extension and the shoulder girdle pulled ventrally. Activity of the abdominal musculature and the flexors at the right hip joint is not needed, and there is little strain on the lumbar spine. The right leg has moved in the vertical sagittal plane of the right hip joint.

Enhancing the stretching of the femoral biceps: The instructions and the movement sequence are as before, but the plane in which the leg is moved is different: it is the vertical plane through the right hip joint that lies parallel to the right shoulder/left hip joint body diagonal. In the starting position, in addition to the flexion at the right hip joint, there is also abduction and external rotation.

243

Enhancing the stretching of the semitendinosus and semimembranosus muscles: The instructions and movement sequence are as before but the plane in which the leg is moved is different: it is vertical and is the plane of the left shoulder/right hip joint body diagonal. In the starting position, in addition to the flexion at the right hip joint, there is also adduction and internal rotation.

These exercises should be performed daily on rising, 10 times each, and again if possible before going to bed. In most cases of moderate to medium ischiocrural shortening, it is possible to achieve a normal range of stretching within 6 weeks. To prevent regression the patient should continue to do the exercises, though not necessarily so frequently.

Notation

'Ischiocrural shortening: R + + +, L +' means that there is extreme ischiocrural shortening on the right and moderate shortening on the left. Additional tests should indicate whether there is a connection between these problems and an asymmetrical antetorsion of the neck of the femur.

Antetorsion and Detorsion

If you lay a femur on a table, its neck points inwards/upwards. In anatomy, we call this 'antetorsion of the femoral shaft.' The angle formed between the axis of the femoral neck and the axis of the femoral condyles is the *angle of antetorsion.* In the newborn, this angle is about 35°, but during development it becomes smaller, being normally about 12° in the fully-grown adult (see Lang and Wachsmuth 1972, Fig. 245). In old age, the angle of antetorsion diminishes further. This normal process of reduction of the angle of antetorsion is also called *detorsion.*

6.4 Postural Statics

Definition: In determining the patient's functional status, the stresses and strains imposed upon a patient's motor apparatus by his posture when upright are assessed under the heading 'statics'.

To understand the functional importance of posture, i. e. the way a person holds his body, we ask ourselves, 'what has to be held up by what?' The answer is, 'The weights of body segments or parts of them have to be held together by passive structures and held up by muscles anywhere where they would otherwise fall down.'

In order better to describe how the intrinsically mobile system of the body is held upright, we speak in terms of building blocks when assessing postural statics, rather than the link-and-chain image we used earlier. Because the human being stands upright, these building blocks have to be arranged one above the other. They are held together by the passive structures of the motor apparatus, on which greater or lesser demands are made, while the muscles, as active structures, prevent the blocks from falling down (Fig. 276).

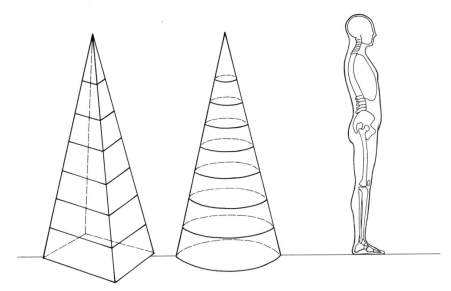

Fig. 276. Pyramid, cone, vertebral column

How can these passive structures which hold the blocks up be protected from strain? What limits the demands made on the muscles?

Building blocks impose least strain on the structures connecting them when their adjoining surfaces are horizontal and congruent and each block is smaller than the one below it – i. e. when they are put together in the form of a pyramid or standing cone. With its triple curvature, the vertebral column fulfils these conditions economically, and its physiological loading thus comes about through vertical compression. Only in the kyphotic region does it have to counteract a constant tendency to fall, due to the preponderance of weight carried ventral to the flexion/extension axis of the thoracic spine; in the lordotic parts of the spine there is a potential perpetuum mobile of equilibrium reactions to any change in the arrangement of the building blocks or the position of the extremities.

What about the substructure of the vertebral column? We have a stable pelvic girdle, which we think of as symmetrical. In two-legged standing, the pelvis, the stable component of the substructure of the vertebral column, balances with its cup-shaped sockets on the spherical heads of the femora. There is therefore no optimum position for the pelvis, only optimum potential mobility at the joints of the hip and lumbar spine.

The thighs stand as if on a transverse cylinder (the femoral condyles), resting on the horizontal tibial plateaux of the lower legs; at the lower end of the latter the prongs of the malleoli of each foot enclose the transverse cylinder of the talus. The talus, as part of the tarsus, articulates with the calcaneus at the subtalar joint, forming the proximal part of the longitudinal arch of the foot, which is twisted distally in the middle foot and together with the phalanges completes the mobile substructure of the vertebral column.

245

Procedure for Assessing the Postural Statics

Starting position: Standing upright. In assessing the postural statics we start at the bottom and work up, examining each level of movement in turn for deviation from the neutral position. It is important to establish whether any particular deviation is due to displacement of a fulcrum or to a change in position of caudal or cranial levers, pointers or gliding bodies, or both. Only deviations need be recorded.

Distribution of Weights in Relation to the Midfrontal Plane and the Plane of Symmetry

The next step is to evaluate the consequences of the deviations in the positions of these joints. These deviations have *altered the distribution of weights.* We study the weight distribution in relation to the midfrontal plane to see which weights have shifted too far forwards (ventrally) or backwards (dorsally), and in relation to the plane of symmetry to see which have shifted too far left or right.

Notation

'+ Weight on heel, knee back/pelvis and abdomen forward/thorax back/head forward in relation to thorax' means: In comparison to economical upright posture there is too much weight on the heel; in consequence, there is increased plantar flexion at the talocrural joint, which has caused the knee to move backwards. Because of hyperextension at the knee joint, the long axis of the thigh has inclined forward, so that the pelvis and the abdomen are too far forward. The thorax has shifted backwards to compensate the weight, and the head has then moved forwards. The midfrontal plane is the bisecting plane with regard to front and back weights. Although the body is in equilibrium, the weights of the body segments are not arranged one above the other as we should like, so the vertebral column is not axially loaded and the posture is uneconomical.

'+ Weight on R foot, GT R/thorax L/head R' means: There is increased weight on the right foot and the greater trochanter has shifted right. The thorax has slipped down to the left and the head is straining away again to the right.

Shearing Stress

The altered distribution of weights can expose the passive structures of the motor apparatus to shearing stress. In notation the following are recorded:

1. Localization of the shearing stress.
2. The weight below the level of stress, giving the direction (apart from downwards) in which that weight is pulling.
3. The weight above the level of stress, giving the direction (apart from downwards) in which that weight is pushing.

Notation

'Lumbar double: From below, the weight of the abdomen exerts a forward/downward pull; from above, the weight of the thorax pushes backwards/downwards.' This means that the lumbar spine is exposed to double shearing stress. In an attempt to correct the patient's statics the weight of the abdomen would have to be brought backwards/upwards and that of the thorax forwards/upwards.

'Shearing stress in lumbar spine: From above, the weight of the thorax is pushing to the right/downwards.' This means that the lumbar spine is exposed to lateral shearing stress. To improve the statics the thorax should be moved left/upwards.

Changes in Muscle Tone

A change in the distribution of weights can also cause *reactive hypotonus* or *reactive hypertonus*. Reactive hypotonus is the result of disengaging muscle activity which, when the statics are normal, is engaged in preventing falling. Reactive hypertonus occurs when muscles which under normal static conditions need not be activated against gravity are called upon to exercise constant fall-preventing activity after a change in the distribution of weights, in order to restore the body's equilibrium.

Notation

'Hypertonus of the upper trapezius, in reaction to the weight of the head' means that, because the head is too far forward in relation to the thorax and the midfrontal plane, the very mobile shoulder girdle has been pulled backwards/upwards by the trapezius to act as a counterweight.

'Hypotonus of the extensors of the thoracic spine, in reaction to the collapse and backward translation of the thorax, with hypertonus of the striated abdominal musculature and the scalene muscles, in reaction to the weight of the thorax' means that the unstable, somewhat hypermobile thoracic spine has lost its dynamic stabilization in the neutral position. It has collapsed and the thorax has slipped backwards in the lower part of the thoracic spine. This causes the weight of the thorax to be suspended from the straight abdominal musculature and the scalene muscles, resulting in hypertonus of these muscles.

'Hypertonus lumbar, paravertebral R, in reaction to elevated pelvis R' means that with the pelvis elevated on the right, the thorax has a tendency to slip down to the left. This tendency activates the lumbar musculature on the right in fall-preventing activity.

Note

Shearing stress on the passive structures is often the cause of pain (periosteal, dystrophic, compressive or radicular symptomatology). Pain is also caused by persistent reactive hypertonus of muscles not designed to be continuously activated; this is ischaemic pain.

Assessment and Notation of Postural Statics as Seen from the Side

Curves in the vertebral column in the plane of symmetry are physiological. The extent of curvature defines postural type. Both emphasized and flattened-out curvatures are considered variants within the norm as long as the economy of the muscle tone is not disturbed. Pathological deviations are characterized by shearing stress on the passive structures of the vertebral column, uneconomical reactive hypertonus or hypotonus of the musculature, and pain due to overstrain.

A more emphasized physiological curvature of the vertebral column reduces the total length of the body; flattening out of the curvature extends it.

The mobility of the vertebral column in the plane of symmetry is distinctly greater in the lordotic than in the kyphotic sections.

Levels of Movement (Fig. 277)

Level 1: Foot/ground = longitudinal arch of the foot (Fig. 278)

+, + +, + + +
−, − −, − − − Longitudinal arches of the feet.

Level 2: Lower leg/foot = talocrural joint (Fig. 279)

+, + +, + + + Dorsiflexion at the talocrural joint.
+, + +, + + + Plantar flexion at the talocrural joint.

Level 3: Thigh/lower leg = knee joint (Fig. 280)

'+, + +, + + + Crus recurvatum' means a bone structure of the lower leg with backward deviation of the proximal third of the lower leg.

'+, + +, + + + FLEX in the knee joint' means that the fulcrum of the joint, the flexion/extension axis of the knee, has been displaced forwards to a greater or lesser extent; the long axis of the lower leg inclines forwards and the long axis of the thigh backwards.

'+, + +, + + + EXT of the knee joint' means genu recurvatum: the fulcrum, the flexion/extension axis of the knee, has been displaced backwards to a greater or lesser extent; the long axis of the lower leg inclines backwards and the long axis of the thigh forwards.

'+ FLEX or + EXT of the lower leg/thigh at the knee joint' means that only the distal/proximal lever at the knee joint shows deviation; for reasons of equilibrium this rarely occurs.

Level 4: Pelvis/thigh = hip joint (Fig. 281)

'+, + +, + + + FLEX at the hip joint' means that the fulcrum, the flexion/extension axis of the hip joint, has been displaced backwards; the long axis of the thigh inclines backwards and the pelvis inclines forwards.

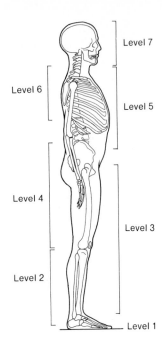

Level 7

Level 6

Level 5

Level 4

Level 3

Level 2

Level 1

Fig. 277. Levels of movement in assessment of postural statics as seen from the side

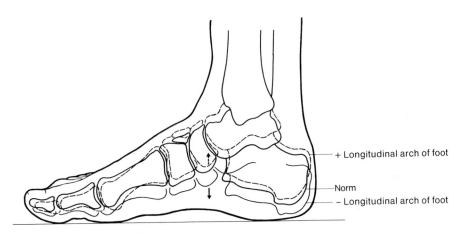

+ Longitudinal arch of foot

Norm

– Longitudinal arch of foot

Fig. 278. Level 1: foot/ground

'+, ++, +++ EXT at the hip joint' means that the fulcrum, the flexion/extension axis of the hip joint, has been displaced forwards; the long axis of the thigh inclines forwards and the pelvis inclines backwards or is upright.

'+ FLEX of the thigh at the hip joint' means that flexion at the hip joint is effected by the distal lever, and therefore the long axis of the thigh inclines backwards.

249

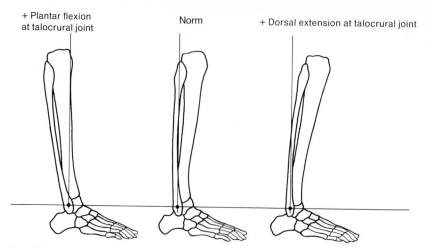

Fig. 279. Level 2: lower leg/foot

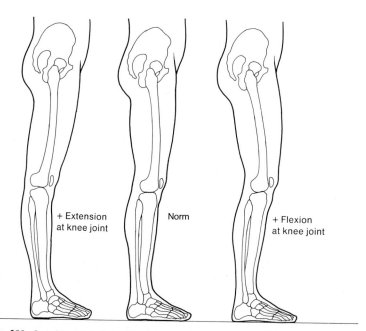

Fig. 280. Level 3: thigh/lower leg

250

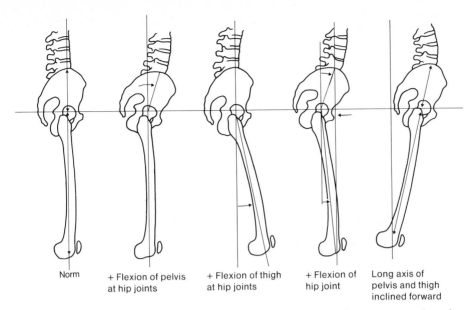

| Norm | + Flexion of pelvis at hip joints | + Flexion of thigh at hip joints | + Flexion of hip joint | Long axis of pelvis and thigh inclined forward |

Fig. 281. Level 4: pelvis/thigh. Only the flexional variants are shown; there are an equal number of extensional variants

'+ EXT of the thigh at the hip joint' means that extension at the hip joint is effected by the distal lever, and therefore the long axis of the thigh inclines forwards.

'+, ++, +++ FLEX of the pelvis at the hip joints' means that flexion at the hip joints is effected by the proximal lever; therefore the pelvis inclines forwards more.

'+, ++, +++ EXT of the pelvis at the hip joints' means that extension in the hip joints is effected by the proximal lever, so the pelvis inclines backwards.

'+, ++, +++ forward inclination of the long axis of the legs and of the pelvis, or of the long axis of the thighs and of the pelvis' means that, with the hip joints approximately in the neutral position, the long axis of the legs and of the pelvis inclines forward due to dorsiflexion at the talocrural joints effected by the proximal lever, or the long axis of the thighs and of the pelvis inclines forwards due to extension in the knee joints effected by the proximal lever.

'+, ++, +++ backward inclination of the long axis of the thighs and of the pelvis' means that, with the hip joints approximately in the neutral position, the long axis of the thighs and of the pelvis inclines backwards due to flexion at the knee joints effected by the proximal lever.

Level 5: The vertebral column (lumbar, thoracic and cervical spine; Fig. 282)

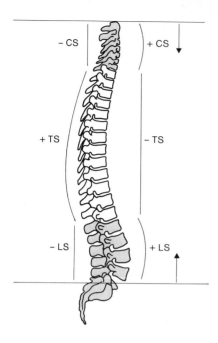

Fig. 282. Level 5: vertebral column. *CS*, cervical spine; *TS*, thoracic spine; *LS*, lumbar spine

'+, + +, + + + LS' means that the lumbar spine is slightly more, considerably more, or very much more than physiologically lordotic.

'−, − −, − − − LS' means that the physiological lordosis of the lumbar spine has been reduced or flattened out or that the lumbar spine is kyphotic.

'+, + +, + + + TS' means that the thoracic spine is slightly more, considerably more, or very much more than physiologically kyphotic.

'−, − −, − − − TS' means that the physiological kyphosis of the thoracic spine has been reduced or flattened out or that the thoracic spine is lordotic.

'+, + +, + + + CS' means that the cervical spine is slightly more, considerably more, or very much more than physiologically lordotic.

'−, − −, − − − CS' means that the physiological lordosis of the cervical spine has been reduced or flattened out or that the cervical spine is kyphotic.

If the physiological curvatures have been displaced, we record the direction in which this has taken place, i.e. lordosis/kyphosis displaced caudally/cranially, and indicate which mobile segments are affected.

'+, + +, + + + Cervical kyphosis'. Kyphosis of the neck affects, roughly, motion segments T3–C5. It is seen in connection with a pronounced thoracic flat back, partic-

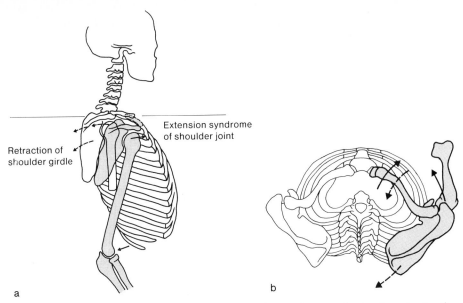

a   b

Fig. 283a, b. Level 6: shoulder girdle/thorax. **a** Extension syndrome and retraction; **b** retraction and protraction of the shoulder girdle

ularly in the midthoracic spine. All motion segments affected by kyphosis of the neck are considerably restricted, tending to immobility in regard to all components of movement.

Level 6: Thorax/shoulder girdle/humerus = sterno- and acromioclavicular joint/ humeroscapular joint (Fig. 283)

'+ Ventral ROT of the clavicle' means that the acromion is ventral and slightly caudal to its physiological position.

'+ Dorsal ROT of the clavicle' means that the acromion is dorsal and slightly cranial to its physiological position ('shoulder girdle retracted').

'+ EXT at the humeroscapular joint' is an *extension syndrome of the humeroscapular joint*. DP acromion is set ventrally/caudally and DP olecranon dorsally/cranially, with extension at the shoulder and flexion at the elbow joint. *Extension syndrome of the humeroscapular joint* is frequently combined with cervical kyphosis. It is accompanied by shortening of the pectoral muscles and triceps brachii.

Level 7: Occiput/atlas, atlas/axis = atlanto-occipital and atlanto-axial joints (Fig. 284)

253

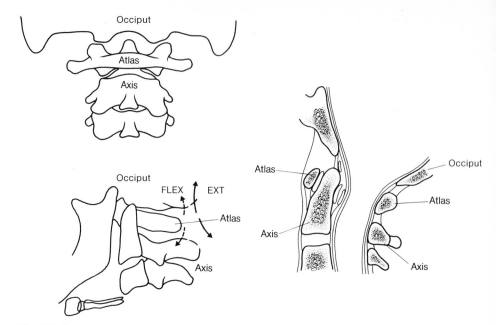

Fig. 284. Level 7: atlas/axis, occiput/atlas

'+, + +, + + + FLEX atlanto-occipital' means that there is a deviation from the neutral position into flexion at the atlanto-occipital and atlanto-axial joints.

'+, + +, + + + EXT atlanto-occipital' means that there is a deviation from the neutral position into extension at the atlanto-occipital and atlanto-axial joints.

These deviations at the uppermost level take place when the curvature of the cervical spine has been altered, in order to allow the glance to be directed forwards again.

● *Example*
Postural statics as seen from the side: Pain-triggering factors caused by shearing stress on the passive structures. Ischaemic pain due to persistent reactive hypertonus of the musculature (Fig. 285).

Notation of Postural Statics as Seen From the Side

● *Example:* Hyperkypholordotic type (normal curves of the back exaggerated)
+ Plantar flexion at the talocrural joints with + EXT at the knee joints with long axes of the pelvis and the thighs inclined forward
LS (lordosis extended cranially to T11)
+ TS (kyphosis extended cranially to C6)
+ CS (from C5)

254

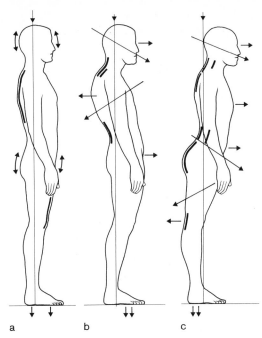

Fig. 285 a–c. Frequent pain-triggering forward/backward deviations from the norm in postural statics. **a** Statics from the side: norm. **b** + Weight on the forefoot. Simple shearing stress in the lumbar spine: the weight of the thorax is pushing backwards/downwards. Double shearing stress in the cervical spine: from below, the weight of the thorax is pulling backwards/downwards; from above, the weight of the head is pushing forwards/downwards. Reactive hypertonus of the musculature: cervical (weight of the head). **c** + Weight on the heel; shearing stress backwards at the knee joint (genu recurvatum); simple shearing stress at the lumbar spine, where the combined weight of the pelvis and abdomen is pulling forwards/downwards from below, and at the cervical spine, where the weight of the head is pushing forwards/downwards from above. Reactive hypertonus of the musculature: gluteal (flexion of the pelvis at the hip joints), lumbar spine (weight of the thorax), and cervical spine (retraction of the shoulder girdle due to weight of the head)

Interpretation: In the assessment of postural statics, the patient's back type is only recorded if it is particularly noticeable or striking. Here, we have genu recurvatum with the flexion/extension axis of the knee joints shifted backwards, reducing flexion at the hip joints. The lumbar lordosis is not increased but extended, which means that the kyphosis of the thoracic spine only begins at T10 and is displaced cranially to C6. The rest of the cervical spine is in increased lordosis. This very frequent phenomenon, which is usually accompanied by kyphosis with stiffness in the upper thoracic and the lower cervical spine, and, in most cases, by increased extension in the rest of the cervical spine, is called a cervical or neck kyphosis. The kyphotic alteration of the lower cervical spine affects the position of the head unfavourably, causing it to be too far ventral to the thoracic spine. The prevailing stiffness causes too great a resistance for the head to be aligned in the long axis of the body.

- *Example:* Flat-back type
- − Longitudinal arch of the feet, more pronounced on the right than on the left
- + Dorsiflexion at the talocrural joints
- + EXT at the hip joints
- − LS/− −TS/−CS
- + EXT at the atlanto-occipital and atlanto-axial joints

Interpretation: This is a case of total flat-back above extended hip joints (forward displacement of the flexion/extension axis of the hip joints). While the lordosis of the lumbar and cervical spine is only slightly reduced, the kyphosis of the thoracic spine has been flattened out, which has very probably resulted in loss of its dynamic stabilization in the neutral position.

Assessment and Notation of Postural Statics as Seen from Behind and In Front

Curvature of the vertebral column in frontal planes is not physiological. On the other hand, the vertebral column is mobile in all its sections in the frontal plane by lateral flexion; this varies from person to person and is most pronounced in the lordotic sections.

Distortion of the vertebral column in transverse planes is not physiological. The mobility of the vertebral column in transverse planes varies from person to person. The functionally important levels of rotation are in the lower thoracic spine, the cervical spine and, with very wide movement tolerance, at the atlanto-occipital and atlanto-axial joints.

Curvature and distortion of the vertebral column accompanied by deformation of vertebrae and loss of mobility are always pathological; these are cases of *structural* scoliotic alteration.

- *Example*

We speak of *scoliotic posture* or *functional scoliosis* when rotation and/or lateral flexion of the vertebral column are compensating lateral tilting and/or rotation of the pelvis. This posture is reversible when the position of the pelvis is corrected. We need to find out whether the tilting or rotation of the pelvis is caused by an anatomical difference in the length of the legs, asymmetry of the pelvis, one-sided valgus deformation at the knee joint, asymmetrical antetorsion angles of the femora or a difference between the angles of the femoral necks. *Structural scoliosis* involves deformity of vertebrae. To differentiate between functional and structural scoliosis we apply the same criteria as in differentiating between functional and structural lumbar kyphosis (see p. 241).

Levels of Movement (Fig. 286)

In analysing the postural statics from behind and in front, we note deviations in the frontal and transverse planes at each level of movement. In frontal planes these deviations are of lateral flexion, abduction and adduction; in transverse planes they are rotational.

256

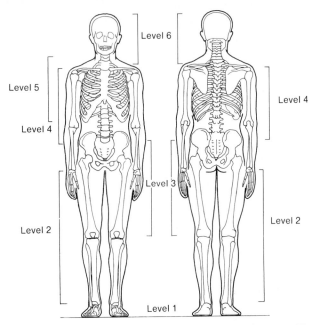

Level 6

Level 5

Level 4

Level 4

Level 3

Level 2

Level 2

Level 2

Level 1

Fig. 286. Diagram of the levels of movement in assessment of postural statics from behind and in front

Level 1: Foot/ground = transverse arch of the foot, inversion/eversion at the subtalar joint, supination/pronation at the talonavicular and tarsometatarsal joints (Fig. 287).

Deviations in the transverse plane

+, + +, + + + Standing forwards/backwards, right/left foot

+, + +, + + + Divergence of the functional long axis of the right/left foot

+, + +, + + + Convergence of the functional long axis of the right/left foot (see p. 265)

+, + +, + + + Hallux valgus, right/left foot

+, + +, + + + Hallux varus, right/left foot

Deviations in the frontal plane

'−, − −, − − − Transverse arches of the feet' means splayed foot.

'+, + +, + + + Supination of the forefoot' means that the distal part of the longitudinal arch of the foot is not fulfilling its arching function.

'+, + +, + + + Pronation of the forefoot' means that the distal part of the longitudinal arch of the foot is overfulfilling its arching function (hollow foot).

257

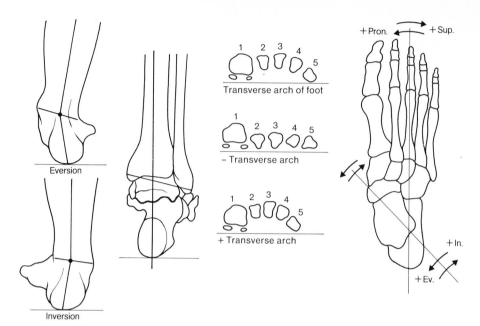

Fig. 287. Level 1: foot/ground

'+, ++, +++ Eversion at the subtalar joint' means that the proximal part of the longitudinal arch of the foot is not fulfilling its arching function.

'+, ++, +++ Inversion at the subtalar joint' means that the proximal part of the longitudinal arch of the foot is overfulfilling its arching function.

The −longitudinal arch of the foot noted in our analysis of the statics from the side view is revealed as sunken arches/flat foot caused by flattening out of the twist of the subtalar plate in +supination and +eversion.

The +longitudinal arch of the foot is revealed as a high instep or hollow foot caused by increased twist of the subtalar plate in +pronation and +inversion.

'+, ++, +++ Track width' means that the distance between the feet, especially the heels, is greater than the distance between the greater trochanters.

'−, −−, −−− Track width' means that the distance between the feet, especially the heels, is less than the distance between the hip joints.

'+, ++, +++ Weight R/L' means the centre of gravity has moved to the right/left.

Level 2: Thigh/lower leg = varus/valgus position of the knee joint, external rotation of the lower leg at the knee joint (Fig. 288).

Deviations in the transverse plane
'+, ++, +++ ER at the knee joint' means the lower leg and the foot are externally rotated at the knee joint, although the latter is in the neutral position in relation to flexion/extension. This is a pathological condition.

258

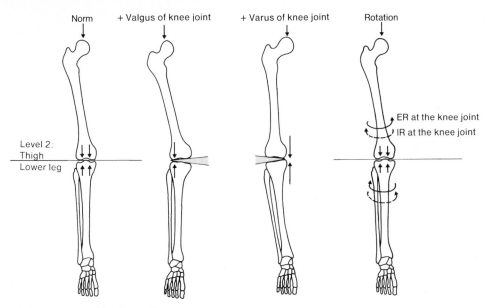

Norm   + Valgus of knee joint   + Varus of knee joint   Rotation

ER at the knee joint
IR at the knee joint

Level 2:
Thigh
Lower leg

Fig. 288. Level 2: thigh/lower leg

Deviations in the frontal plane
'+, ++, +++ Varus of the knee joint' (bow legs) means that the knee joints cannot be brought together medially even when the feet are touching medially.

'+, ++, +++ Valgus of the knee joint' (knock knees) means that, with a normal track width, the knee joints touch medially.

'+, ++, +++ Varus of the lower leg' means that the lower leg has a varus bone structure, the proximal third exhibiting valgus position.

> Level 3: Pelvis (line connecting right and left iliac spines)/thigh = internal rotation/ external rotation, abduction/adduction (Fig. 289).

Deviations in the transverse plane
'+, ++, +++ Medial rotation of the FLEX/EXT axis of the R/L knee joint' means the hip joint exhibits internal rotation effected by the distal pointer.

'+, ++, +++ Lateral rotation of the FLEX/EXT axis at the R/L knee joint' means the hip joint exhibits external rotation effected by the distal pointer.

'IR/ER of the pelvis at the R/L hip joint' means that internal/external rotation has been effected at the hip joint by the proximal lever.

'Medial rotation of the FLEX/EXT axis at the L knee joint with simultaneous lateral rotation of the R knee joint and +ROT of the pelvis in relation to the midfrontal

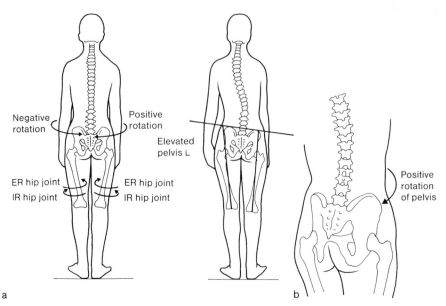

Labels in figure:
Negative rotation — Positive rotation — Elevated pelvis L — ER hip joint — IR hip joint — ER hip joint — IR hip joint — Positive rotation of pelvis

a   b

Fig. 289a. Level 3: pelvis/thigh. **b** positive rotation of pelvis and + flexion at the right hip joint

plane' means that the knee joints and the pelvis have both rotated out of the midfrontal plane; the hip joints may have remained in the neutral position.

Deviations in the frontal plane
'+, + +, + + + Elevated pelvis R' means the pelvic crest on the right is higher than that on the left, i. e. there is adduction at the right hip joint and abduction at the left. If the fulcrum has been displaced to the right, there is adduction at the right hip joint and abduction at the left hip joint effected by both levers, or the other way round if the fulcrum has been displaced to the left.
In pelvic scoliosis, viewing from behind a pelvis elevated on the right, the right side of the pelvis may appear narrow and high and the left side broad and low; viewing from in front, the reverse is true. This is notated as: '+, + +, + + + Pelvic scoliosis: from behind, R narrow/high, L broad/low; from in front, the reverse.'
A pelvis elevated on the right may also indicate that the right side of the pelvis is larger overall than the left side. If a pelvis is still elevated in the sitting position, the case is one of an asymmetrical pelvis, not of a difference in leg length. It is advisable in such a case to use padding to balance the lower side of the pelvis during prolonged periods of sitting.

Elevated pelvis caused by functional differences in leg length: One-sided − − longitudinal arch of the foot, +eversion at the subtalar joint, +FLEX/+EXT at one knee joint, +ABD/ADD/FLEX/EXT at one hip joint.

Elevated pelvis caused by anatomical differences in leg length: Differences in length between the diaphyses of the thigh and the lower leg after fractures or operations, or con-

260

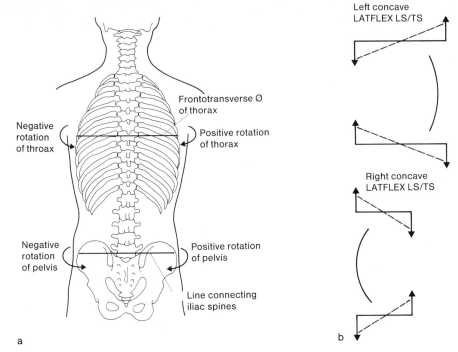

Left concave
LATFLEX LS/TS

Right concave
LATFLEX LS/TS

a

b

Fig. 290 a, b. Level 4: lumbar and thoracic spine

stitutional. Varus/recurvatum of lower leg, +varus, +valgus at one knee joint, +varus, +valgus at femoral neck, +size of R/L side of pelvis.

Level 4: Position of the lumbar/thoracic spine and shape of the thorax, lateroflexion/rotation/torsion/translation of the lumbar/thoracic spine (Fig. 290).

Deviations in the transverse plane
'Pelvis/thorax + / − ROT' means that the pelvis with the line connecting the right and left iliac spines as pointer, or the thorax with its frontotransverse diameter as pointer, has performed positive/negative rotation in relation to the midfrontal plane.

'Pelvis +ROT, thorax −ROT' means that the line connecting the iliac spines has performed positive rotation and the frontotransverse diameter of thorax negative rotation in relation to the midfrontal plane. If there is a difference between the extent of rotation of the pelvis and of the thorax, the pointer with the lesser rotation is given a + or − sign and the pointer with the greater rotation two + + or two − − signs.

'Pelvis +ROT, thorax + +ROT' means that BSs pelvis and thorax have performed positive rotation, the thorax more than the pelvis: a positive rotation has taken place between the pelvis and the thorax.

261

'Scoliosis of LS/TS: Lumbar bulge L, costal bulge centre of thorax R' means that a structural scoliosis is present, with a probable tendency at the lumbar bulge to right concave lateral flexion and negative rotation of the pelvis, and at the costal bulge to left concave lateral flexion and positive rotation of the thorax.

Deviations in the frontal plane
'Elevated pelvis R with compensatory translation of the thorax to R' means that the elevated pelvis has forced the lumbar spine to adapt in right concave lateral flexion while the thorax shifts in a translatory movement over the raised side of the pelvis. If the sole of the left shoe were built up, the lateral flexion and translation would disappear.

'Elevated pelvis R, R concave LATFLEX of the vertebral column from L5 to T5 with L thorax slipping downwards to the left; translation of the head R' means that with a pelvis standing high on the right, a widely arching right concave lateral flexion of the lumbar spine and half the thoracic spine causes the weight of the thorax to move so far left that it slips left in a translatory movement. The head is translated to the side on which the pelvis is raised, an indication that the right foot is carrying more weight than the left. Reactive hypertonus of the right lateral lumbar musculature is possible.

Level 5: Shoulder girdle and arms/thorax (Fig. 291).

Deviations in the transverse plane
'R shoulder forward/back' means that the right shoulder joint is in front of/behind the midfrontal plane and the medial border of the scapula is correspondingly further from/closer to the spinous processes of the thoracic spine than on the opposite side.

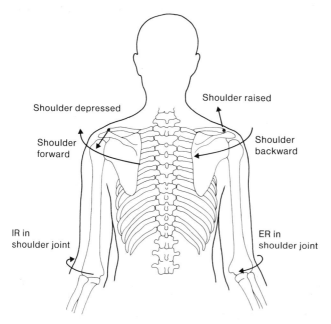

Fig. 291. Level 5: shoulder girdle/arms/thorax

If both shoulders are forward of or behind the midfrontal plane, we speak of protraction/retraction of the shoulder girdle (Fig. 283 b).

'Thorax +ROT with R shoulder forward' means that the shoulder girdle has performed a negative rotation and the thorax a positive rotation in relation to the midfrontal plane; the thorax has of course also rotated in relation to the pelvis and the head.

The forward/backward position of the shoulder could also be caused by asymmetry or scoliosis of the thorax or asymmetry of the right or left clavicle. It is not difficult to see what the real cause is.

Deviations in the frontal plane
'High/low shoulder R' means that the right shoulder joint is higher/lower than the left.

A pelvis elevated on the right with a low shoulder on the same side indicates right concave lateral flexion of the lumbar and thoracic spine.
A pelvis elevated on the right with a high shoulder on the same side indicates compensatory translation of the thorax towards the raised side of the pelvis.

> Level 6: Head/atlanto-occipital and atlanto-axial joints/cervical spine: Occiput/atlas/axis, lateral flexion/rotation/torsion/translation (Fig. 292).

In two-legged standing the head is normally over the middle of the support area. Seen from the side, it is over the feet (see p. 219). If only one leg is weight-bearing, the head should be over that leg. One should therefore never attempt to correct the statics of standing through a translation of the head, since this is already in the right place. One should try to bring about the change reactively, without altering the distribution of

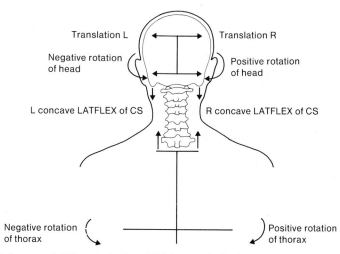

Fig. 292. Level 6: head/atlanto-occipital and atlanto-axial joints cervical spine

weight over the feet, by guiding a distance point on BS pelvis in a certain direction in economical activity and observing the reactive movements of BS thorax, or vice versa.

Deviations in the transverse plane
Positive/negative rotation of the head is rare and indicates a limitation of the field of vision or a contracture of the sternocleidomastoid muscle.
Translation of the head forwards and to the right/left indicates that weight transference backwards and to the left/right has taken place in an area lower down. If the head is over the feet, the displacement of the head should be registered in relation to the thorax.
For rotation and torsion, the cervical spine should be compared with lower sections of the vertebral column and interpreted in relation to them.

Deviations in the frontal plane
Translation of the head to the right/left with the eyes looking to the front cannot be meaningfully interpreted until we know whether there has been weight transference to the left/right lower down, at the level of the pelvis or thorax. Lateroflexional deviations of the cervical spine should also be assessed in relation to the lower sections of the vertebral column.

Remark: As the head is the uppermost weight, no further weight transference is possible above it, so if it deviates out of the long axis of the body, reactive hypertonus in the level of movement below it necessarily results. We have already indicated that the head's position over the feet with the eyes looking forwards is in a sense a constant, but nevertheless it needs to be understood and interpreted in relation to the postural statics as a whole.

Notation of Postural Statics as Seen from Behind and In Front

- *Example* (Fig. 293)
+ Inversion L subtalar joint, + eversion R subtalar joint, standing with the weight over the hindfoot and + weight on L foot.
Slightly elevated pelvis L with thorax slipping downwards R.
Abduction syndrome at L shoulder joint with slightly high shoulder.
Slight L translation of head.
Distribution of weight in relation to the midfrontal plane/plane of symmetry: + weight L, pelvis L/thorax R/head L.
Muscle tonus: Hypertonus paravertebral lumbar L, reactive to weight of thorax.

Interpretation: The elevated pelvis on the left with increased weight placed upon the left foot has led the thorax to slip downwards to the right, because at the level of the thorax the weight was not moved to the left early enough to compensate. Reactive lumbar hypertonus occurs, very probably causing ischaemic pain.

- For further *examples* see pp. 288 ff.

264

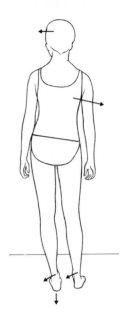

Fig. 293

6.5 Basic Gait Test

We have characterized the legs as the mobile substructure of the vertebral column. It is obvious, therefore, that all disorders of gait and all asymmetries or non-functional loading of the legs play an important role in motor behaviour and in all syndromes affecting the vertebral column. In determining the functional status, a comparative assessment of the axes of the legs and a basic gait test should not be omitted. For an exhaustive functional analysis of gait and gait training, the reader is referred to *Gangschulung* (Klein-Vogelbach 1990).

Note

For the legs to bear weight well, it must be possible to align the talocrural, knee and hip joints economically above each other in a vertical plane.

For economical walking, it must be possible to arrange the flexion/extension axes of the hip and knee joints and those of the proximal joint of the large toe parallel to each other; the longitudinal arches of the feet must be sufficiently flexible to allow the *functional long axis of the foot* to point forwards, while the *anatomical long axis* shows a physiological divergence of about 11°.

Definitions: The *anatomical long axis of the foot* runs from the middle of the back of the heel through the proximal phalangeal joint of the second toe (Fig. 294).

The *pronation/supination axis* runs from the middle of the back of the heel through the proximal phalangeal joint of the third toe (Fig. 294).

265

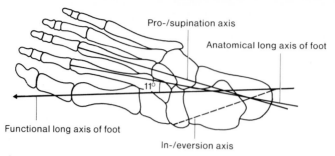

Pro-/supination axis

Anatomical long axis of foot

11°

Functional long axis of foot

In-/eversion axis

Fig. 294. Anatomical long axis of the foot, pronation/supination axis, inversion/eversion axis and functional long axis of the foot

The *inversion/eversion axis* runs from ventral-medial-cranial (navicular-talus) to ventral-lateral-caudal (calcaneus) (Fig. 294).

The *functional long axis of the foot* runs from the tubercle of the lateral calcaneal tuberosity to the centre of the proximal phalangeal joint of the large toe. In normal gait it points forwards in the direction of progress (Fig. 294).

Since the anatomical long axis of the foot is at right angles to the flexion/extension axis of the talocrural joint, it forms an angle of about 11° with the functional long axis of the foot. This angle corresponds to the physiological divergence of the anatomical long axis of the foot from the direction of progress.

In the normal roll-on action of the foot, pronation of the forefoot towards the ground occurs when the heel leaves the ground. At this moment the continuing movement of the free leg, with the critical internal rotation of the thigh at the knee joint of the weight-bearing leg, is actively buttressed by internal rotational activity of the lower leg at the knee joint (see *Gangschulung*). The weight transfer during the roll-on of the foot goes straight forwards to the proximal joint of the large toe. If the weight transfer occurs too early and goes medially, it causes eversion at the subtalar joint; if it goes laterally it causes inversion at the subtalar joint and supination at the tarsometatarsal joints. In these cases the patient moves his knee joint backwards by extension at the knee joint and often rotates the knee joint medially by internal rotation at the hip joint. This is a typical limping pattern (Fig. 295).

Examining the Axes of the Non-Weight-Bearing Leg

Examination in Frontal Planes

Starting position: Lying supine. The hip joint of the leg to be examined is in the neutral position. As an aid to measuring we use the line connecting the centre of the hip joint and the patella. This we call the *varus/valgus line* and we observe how the projection of this line passes through the talocrural joint. *Norm:* The varus/valgus line passes through the centre of the talocrural joint (Fig. 296).

266

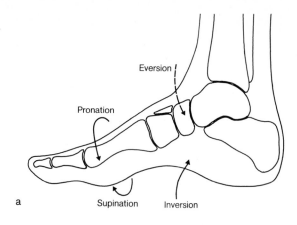

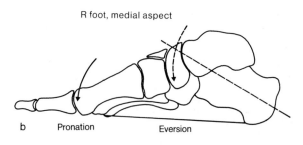

R foot, medial aspect

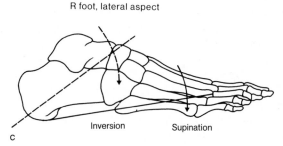

R foot, lateral aspect

Fig. 295 a–c. Definition of inversion/eversion and pronation/supination in roll-on activity of the foot

Valgus deviation at the knee joint: The varus/valgus line passes medial to the centre of the talocrural joint. If it meets the medial malleolus, there is a minor valgus deviation at the knee joint. If it passes medial to the malleolus, the distance (in cm) should be recorded.

Varus deviation at the knee joint: The varus/valgus line passes lateral to the centre of the talocrural joint. If it meets the lateral malleolus, there is a minor varus deviation at

267

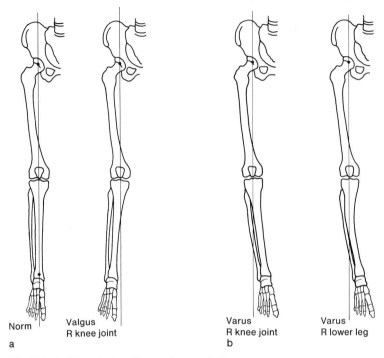

Norm	Valgus		Varus	Varus
a	R knee joint		R knee joint	R lower leg
			b	

Fig. 296 a, b. Varus/valgus line (patient supine)

the knee joint. If it passes lateral to the malleolus, the distance (in cm) should be recorded.

Varus deviation of the lower leg: This is present when the proximal third of the tibia exhibits outward angulation from the knee joint and the distal two-thirds inward angulation. The tibia is curved, the convexity facing outwards. If the varus/valgus line meets the centre of the talocrural joint, the case is one of varus of the lower leg (crural varus) within the norm; if it meets the lateral malleolus or passes further laterally, the distance (in cm) should be recorded.

Note

Where there is more or less pronounced valgus deviation at the knee joint, the foot makes contact with the ground through its medial border. To achieve contact between the sole of the foot and the ground, adaptive inversion at the subtalar joint and slight supination of the forefoot are needed.

In the case of a more or less pronounced varus deviation at the knee joint or the lower leg, the foot makes contact with the ground through its lateral border. To achieve contact between the sole of the foot and the ground, adaptive pronation of the forefoot and slight eversion at the subtalar joint are needed.

268

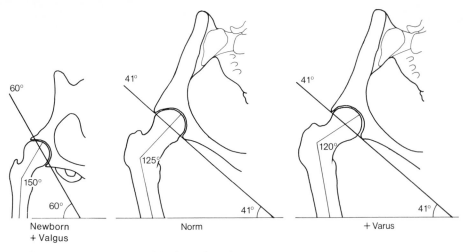

Fig. 297. + Varus/+ valgus of the femoral neck

To estimate varus or valgus deformation of the femoral neck, we consider the varus/ valgus line as one side of a triangle with the greater trochanter at its apex. If the tri- angle is low and the angle is wide, there is valgus deformation of the femoral neck; if the triangle is high and the angle small, there is varus deformation of the femoral neck (Fig. 297).

Remark: If the greater trochanters are constitutionally abnormally close or far apart, this also indicates varus or valgus deformation of the femoral neck.

Examination in Sagittal Planes

Over-extension of the knee joint (genu recurvatum), causing the convex aspect of the axis of the leg to face backwards, must be identified when assessing the long axis of the leg. In genu recurvatum the long axis of the lower leg inclines backwards and the talo- crural joint is in slight plantar flexion, while the long axis of the thigh inclines forwards with slight extension at the knee and hip joints: the flexion-inhibiting activity of the quadriceps ceases to function (Fig. 298).
A bony deformation of the lower leg in which the convex aspect faces backwards (crus recurvatum) disrupts the normal distribution of weight along the axes of the legs inso- far as the knee joint changes its position relative to the talocrural joint, and in the su- pine position the thigh is slightly flexed at the hip joint (Fig. 299).
These two possible conditions should be borne in mind if, in assessing the postural statics, increased plantar flexion is noticed at the talocrural joint.

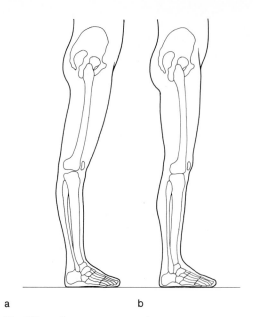

a   b

Fig. 298. a Genu recurvatum; **b** norm

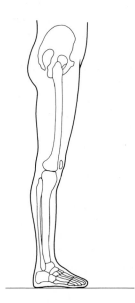

Fig. 299. Crus recurvatum

Tibial torsion is examined with the patient lying supine, particular attention being paid to whether it is symmetrical on the left and the right leg. With one hand we grasp the transverse axis of the head of the tibia and set it parallel to the base support, the tibial tuberosity facing laterally, while the other hand palpates the points of the lateral and medial malleoli, enabling us to judge the extent of torsion of the distal end of the tibia. Normal torsion is 23°. Lower values are recorded as '−, − −, − − − Tibial torsion', higher values as '+, + +, + + + Tibial torsion.'

Note

Increased tibial torsion suggests that the amount of torsion present is greater than the reduction that has taken place of the natal antetorsion. If this is the case, the test of rotation at the hip joints will show increased internal rotation of the neck of the femur. As the transverse axis of the femoral condyles is usually arranged fronto-transversely at the start of this examination (see Debrunner 1971), external rotation has already taken place at the hip joint.

The norm of torsion at the distal end of the tibia is 23°, the maximum 48° and the minimum 0° (Lang and Wachsmuth 1972).
The norm of antetorsion at the proximal end of the femur is 12°, the maximum 37°, the minimum 25° retrotorsion (Lang and Wachsmuth 1972). The extent of antetorsion can be seen by the extent of medial rotation of the transverse axis of the femoral condyles in standing.
With great antetorsion, an increased tibial torsion can correct the forward orientation of the functional long axis of the foot but not the medial rotation of the knee joint.
With great retrotorsion, a reduced tibial torsion can correct the forward orientation of the functional long axis of the foot but not the lateral rotation of the knee joint.
Great antetorsion and reduced tibial torsion deflect the forward orientation of the functional long axis of the foot medially. The position of the knee joint in relation to the functional long axis of the foot is better but problems arise in walking.
With great retrotorsion and tibial torsion, the forward orientation of the functional long axis of the foot is deflected laterally. The position of the knee joint in relation to the functional long axis of the foot is better but problems arise in walking.

Note
In all the following illustrations the axis of the neck of the femur is in the frontal plane of the hip joint.

Fig. 300: Norm: 12° antetorsion, 23° tibial torsion; the functional long axis of the foot points forwards.
Fig. 300 a_1: 4° Antetorsion, 18° tibial torsion; the functional long axis of the foot is deflected laterally by about 3°.

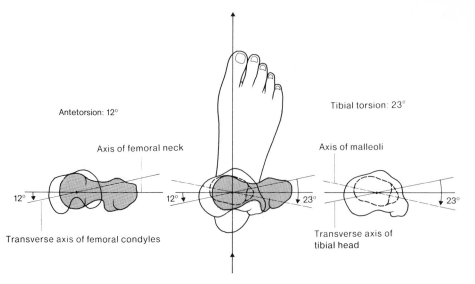

Antetorsion: 12°

Axis of femoral neck

Tibial torsion: 23°

Axis of malleoli

12°

12°

23°

23°

Transverse axis of femoral condyles

Transverse axis of tibial head

Fig. 300

Fig. 300 a₂: 37° Antetorsion, 48° tibial torsion; the functional long axis of the foot points forwards; the knee joint is medially rotated 25° further than normal.

Fig. 300 b₁: − 12° Retrotorsion, 10° tibial torsion; the functional long axis of the foot is deflected laterally by about 11°; the knee joint is laterally rotated.

Fig. 300 b₂: 20° Antetorsion, 28° tibial torsion; the functional long axis of the foot is deflected medially by about 3°; the knee joint is medially rotated 8° further than normal.

Fig. 300 c₁: 4° Antetorsion, 30° tibial torsion; the functional long axis of the foot is deflected laterally by about 15° because the reduced antetorsion and increased tibial torsion compound each other.

Fig. 300 c₂: 20° Antetorsion, 13° tibial torsion; the functional long axis of the foot is deflected medially by about 18° because the increased antetorsion and reduced tibial torsion compound each other.

This attempt to interpret the contrary transverse torsions of the distal end of the tibia and of the transverse axis of the femoral condyles is of considerable help in the basic gait test. We assume that the functional long axis of the foot and the flexion/extension axis of the talocrural joint form a ventrolateral angle of about 101°. This corresponds to the right angle formed by the anatomical long axis of the foot and the flexion/extension axis of the talocrural joint, plus about 11° physiological divergence. In our observation we assess the axis of the malleoli, i.e. the line from the lateral to the medial malleolus, which can be seen quite plainly and is almost parallel to the flexion/extension axis of the talocrural joint. This allows us to tell which patients are able to orient the functional long axis of their foot forwards when walking and why others are unable to do this.

272

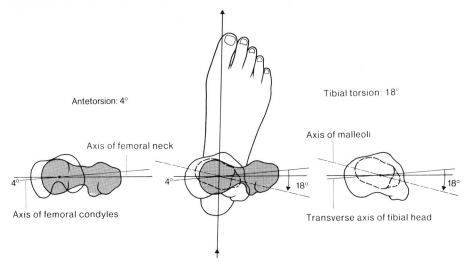

Fig. 300 a₁

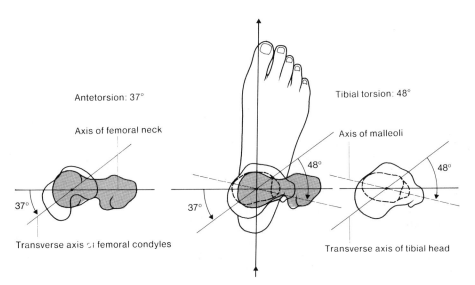

Fig. 300 a₂

We must also determine by how much the flexion/extension axes of the knee joints deviate laterally/medically in relation to the functional long axis of the foot. If the ligamentous apparatus is sound, the transverse axes of the femoral condyles and of the tibial head are almost parallel. We assume that the axis of the neck of the femur as the axis of reference is always in the midfrontal plane. Antetorsion of the femoral neck shows as medial rotation of the transverse axis of the femoral condyles, retrotorsion as lateral rotation, and these determine the position of the flexion/extension axis of the knee joint. The position of the transverse axis of the tibia (the axis of reference for the

273

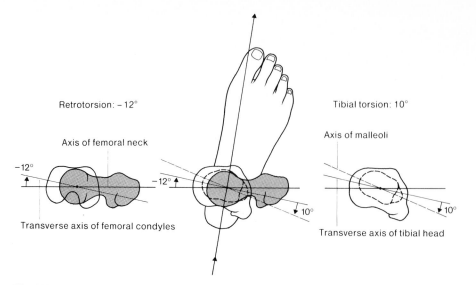

Retrotorsion: – 12°

Axis of femoral neck

−12°

−12°

Transverse axis of femoral condyles

Tibial torsion: 10°

Axis of malleoli

10°

Transverse axis of tibial head

Fig. 300 b₁

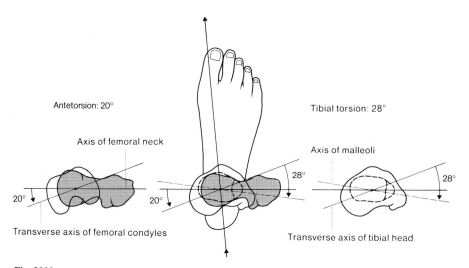

Antetorsion: 20°

Axis of femoral neck

20°

20°

Transverse axis of femoral condyles

Tibial torsion: 28°

Axis of malleoli

28°

28°

Transverse axis of tibial head

Fig. 300 b₂

axis of the malleoli, by which lateral rotation of the distal end of the tibia is measured), being almost parallel to the transverse axis of the femoral condyles, is accordingly dictated by the position of the latter. Constant deviation at a knee joint in the neutral position takes its toll of the ligamentous apparatus. These torsions are developed during longitudinal growth periods, as is clearly shown by a comparison between fetal and adult skeletons (Fig. 301).

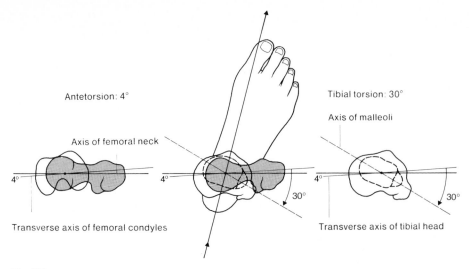

Antetorsion: 4°

Axis of femoral neck

Tibial torsion: 30°

Axis of malleoli

4°   4°   4°

30°   30°

Transverse axis of femoral condyles

Transverse axis of tibial head

Fig. 300 c₁

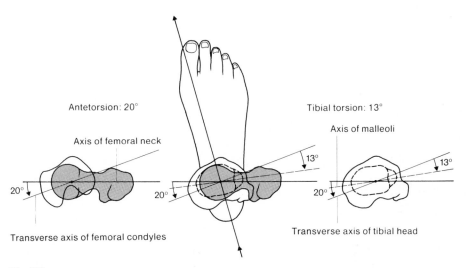

Antetorsion: 20°

Axis of femoral neck

Tibial torsion: 13°

Axis of malleoli

13°   13°

20°   20°   20°

Transverse axis of femoral condyles

Transverse axis of tibial head

Fig. 300 c₂

Walking on the Spot

Starting position: Standing on two legs. The feet are in line with the hip joints and the functional long axes of the feet point forwards. During alternating pressure activity on the soles of the right and left feet, with each leg alternately supporting and in free play, we observe by watching and palpating whether:

275

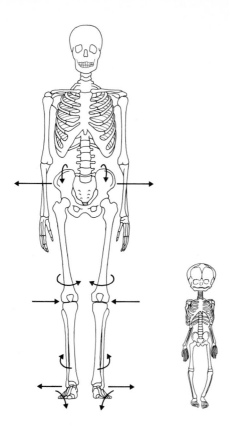

Fig. 301. Adult and fetal skeleton: compare the statics of the axes of the legs

- In the starting position the norm prevails (i.e. no hypertonus) or whether there is lumbar hypertonus working against gravity.
- The muscle tone in the lumbar region alternates normally when the patient walks on the spot. When the weight is being supported on one leg, we expect reactive lumbar tone only on the side of the leg in free play, indicating that the leg is laterally suspended from the pelvis, which in turn is suspended from the vertebral column and BS thorax.

If we find lumbar tone directed against gravity, occurring both in the starting position and on the side of the supporting leg as weight is transferred from side to side, this is recorded as 'persistent lumbar hypertonus'; it tells us that the pelvis has lost its potential mobility at the hip jonts and the joints of the lumbar spine.

Anchoring the Pelvis to the Supporting Leg

Transferring load from right to left, standing on the spot. We observe whether:

- The line connecting the iliac spines remains horizontal.
- The line connecting the iliac spines is dropped or elevated on the side of the leg in free play.

276

- The greater trochanter moves medially, laterally or dorsally, or remains stationary.
- The side of the free play leg rotates dorsally.

We note whether the pelvis is anchored to the supporting leg in adduction, abduction, flexion, or external rotation controlled by the abductors.

Step Mechanism

The patient is asked to walk at normal speed (about 120 steps/min) with or without shoes and the physiotherapist observes:

- Are the steps active or reactive?
- How long are the steps?
- Is the track width normal?
- Is there a normal roll-on phase? (Fig. 302, Table 1.) If not, is the functional long axis of the foot oriented medially or laterally?

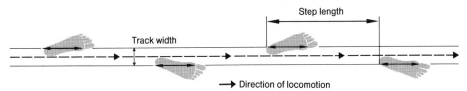

Step length

Track width

→ Direction of locomotion

Fig. 302. The track width is a constant. In normal gait it is always narrower than the width between the hip joints: otherwise, during the rolling-on of the weight-bearing foot on the floor, the internal rotation of the pelvis with the free play leg would not take place at the weight-bearing hip joint, nor would the continuing lateral rotation of the femoral condyles take place internal-rotationally in the weight-bearing knee. The normal track has the narrowest width at which the feet do not get in each other's way during walking. In that way, with each step there is the most movement forwards and the least movement sideways

Table 1. Co-ordination in time of the supporting leg/free play leg phase (S/FP), the bipedal floor contact phase (TWO), and the roll-on the foot on the floor along the functional long axis of the foot

Right			Left	
	Heel contact	TWO		Toes lift off
Roll-on phase	Forefoot contact	S/FP	Free play phase	
	Toes lift off	TWO		Heel contact
Free play phase		S/FP	Roll-on phase	Forefoot contact
	Heel contact	TWO		Toes lift off
Roll-on phase	Forefoot contact	S/FP	Free play phase	
	Toes lift off	TWO		Heel contact
Free play phase		S/FP	Roll-on phase	Forefoot contact
	Heel contact	TWO		Toes lift off
Roll-on phase	Forefoot contact	S/FP	Free play phase	
	Toes lift off	TWO		Heel contact

- What is the position of the flexion/extension axis of the knee joint in the weight-bearing phase? Does weight-bearing occur with the foot inverted or everted?
- In the weight-bearing phase, does the greater trochanter move backwards in relation to the point of contact between the supporting foot and the ground?
- Is the frontotransverse diameter of the thorax at right angles to the direction of progress or does it rotate backwards on the side of the supporting leg at mid-stance?
- Is the arm swing normal or is the swing restricted on one or both sides to the forearm?
- Does the counterswing of the arms occur in parallel planes? Or is the arm swing diagonal? If so, does the arm swing move from behind/lateral to in front/medial or from behind/medial to in front/lateral?

Notation

'Active steps': This means that the normal reactive step mechanism is disturbed. This mechanism can function properly only if, during the stance phase, the whole of the weight of the body above the greater trochanter is brought forwards over the supporting leg.

'Absence of normal roll-on phase': This can be observed when at mid-stance the heel cannot be raised immediately it has made contact with the ground. There may be a number of causes for this:

- If the anatomical long axis of the foot is too divergent, the inversion/eversion axis of the subtalar joint moves into a vertical plane pointing forwards in the stance phase, and the ensuing roll-on is effected through eversion. The torsion of the 'plate' of the subtalar foot disappears and thus so does the longitudinal arch of the foot; the extent of the roll-on is severely curtailed.
- If the physiological divergence of the anatomical long axis of the foot disappears during walking and the axis points forwards, the functional long axis of the foot converges and the foot points medially. In attempting the roll-on phase, transferral of the weight from the lateral border of the foot is delayed. The foot tends to tip over laterally, and pronation of the forefoot cannot take place.
- If the flexion/extension axis of the knee joint in the stance phase is medially rotated (frequently observed with increased divergence of the long axis of the foot) or laterally rotated (observed in cases of significantly reduced tibial torsion and where the hip joints are of the external rotation type), the roll-on phase cannot be performed correctly.

Backward movement of the greater trochanter in the stance phase can be observed when flexion at the hip joint is increased by backward displacement of the fulcrum as the weight on the leg increases.

Rotation of the frontotransverse diameter of the thorax in its horizontal transverse plane means that the rotation tolerance in the thoracic spine is exhausted by the movement excursion of the cranial pointer. This means that the pelvis is no longer capable of internal rotation at the hip joint and the thigh in the knee joint of the supporting leg and thus cannot bring the free play leg forwards. Normal arm swing is thus disrupted. This disruption should be recorded in terms of its origin, i.e. as a rotation of the frontotransverse diameter of thorax.

6.6 Motor Behaviour in Bending Down

We bend over or work with bent posture to enable our hands or a tool held in our hands to reach the ground or a work surface. In such activities it is often important to see what the hands are doing, to see the thing that they are trying to grasp, or to watch the movements of the tool which they are holding. When examining motor behaviour in bending down, we are not initially concerned with the question of whether the patient is able to lift a load without causing himself an injury; we are rather concerned to establish whether his back is capable of normal function. The behaviour of his back will, however, decide whether the patient will be able to lift a load without incurring injury.

> **Note**
> If a patient's bending behaviour is poor, whether due to his condition or his constitution, he will not be able to learn how to lift weights without damaging his motor apparatus.

For sound bending behaviour, BSs pelvis, thorax and head must remain aligned in the long axis of the body. This means that when the long axis of the body has to be bent over forwards, the fall-preventing activities of the musculature of the vertebral column must be those of extensional stabilization. The lumbosacral junction in particular must be extensionally stabilized. The strain imposed by BSs pelvis, thorax and head (which are stabilized together) upon the musculature increases the further they incline forward, and is at its greatest when the long axis of the body reaches the horizontal. As soon as the forwards-bending, stabilized long axis of the body brings too much weight forwards, the lumbar spine automatically destabilizes and flexes and the pelvis extends in the hip joints, so as to maintain equilibrium during the constant-location movement of bending forwards by displacing weight backwards.

> **Note**
> We differentiate between a *vertical* and *horizontal type* in bending behaviour. On the whole, the vertical type is seen in people whose upper length is constitutionally the longer, particularly if it is also top-heavy (main weight in the cranial part of the thorax, the shoulder girdle, arms and head; Fig. 303). The horizontal type is seen in people whose lower length is the longer, particularly if the thighs are long and the main weight is in the pelvis, abdomen and thighs (Fig. 304).

If we define bending the back as a shortening of the total length of the body, and see the stabilized BSs pelvis, thorax and head as a single lever, we have four levels of movement at which the overall body length can be reduced by moving weights backwards or forwards (Fig. 305).
Level 1 is the point of contact between the foot and the ground. The proximal joints of the toes are the fulcra about which weight can be transferred forwards. Proximal DP heel moves forward/upward; there is extension at the proximal joints of the toes, effected by the proximal lever.

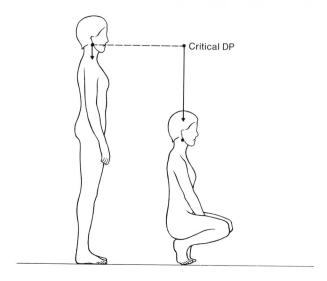

Fig. 303. Vertical bending type

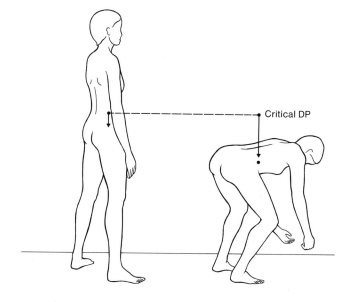

Fig. 304. Horizontal bending type

Level 2 is the talocrural joint, where dorsiflexion can transfer weight forwards or plantar flexion backwards. The moving proximal distance point is the patella, which moves forwards/backwards with dorsiflexion/plantar flexion at the talocrural joint effected by the proximal lever, with flexion/extension at the knee joint through displacement of the fulcrum.

Level 3 is the knee joint, where flexion effected by the proximal lever can transfer weight backwards. The heel is the distal distance point, the greater trochanter is the proximal distance point.

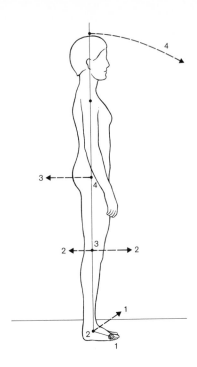

Fig. 305. The four levels of movement at which
the overall length of the body can be reduced

Level 4 is the hip joint, where weight can be transferred forwards by flexion at the hip
joint effected by the proximal lever. The patella is the distal distance point, the vertex
the proximal distance point.

Note

The reactive arm swing in symmetrical roll-on along the functional long axes of the
feet clearly shows the relation of upper to lower length and the distribution of
weights.

If the lower length is greater and there is more weight in the pelvis, a large arm
swing will result, because the forward/backward displacement of critical DPs greater trochanters (flexion/extension axes of the hip joints) results in a great deal of
weight being transferred, eliciting activated passive buttressing in the form of a correspondingly large arm swing (Fig. 306).

If the lower length is shorter and there is more weight in the thorax, shoulder girdle
and head, a small arm swing results, because the transfer of BSs thorax and head
brings enough weight forwards/backwards to necessitate only a small forward/
backward displacement of critical DPs greater trochanters (flexion/extension axes
of the hip joints), and therefore only a small arm swing or none at all is required as
an equilibrium reaction (Fig. 307).

281

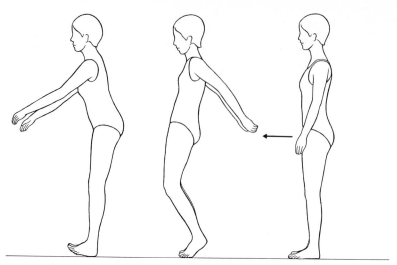

Fig. 306. Reactive arm swing in a person with + lower length

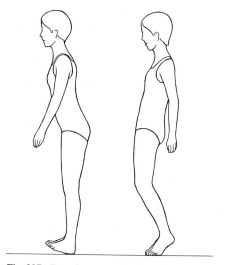

Fig. 307. Reactive arm swing in a person with − lower length

Vertical Bending Type

If assessment of the relationships of lengths and the distribution of weights in a patient suggests that he may be of the vertical bending type, we give the following instruction: 'Let both your knees move forwards/downwards and the heels lose contact with the ground, while your suprasternal notch moves vertically downwards.'

Analysis: Because of the greater length and top-heaviness of the upper length, the long axis of the body should remain more or less vertical. In order that sufficient counter-

282

weight be transferred forwards, the body's centre of gravity is moved forwards within the support area (the heels lose contact with the ground and DP patella moves forwards/downwards). This causes the weight moved backwards by the fulcrum at the knee joint to be well distributed over the support area. Since the thighs are short, the braking activity of the quadriceps is not too high. This type of bending very much depends on sound condition of the quadriceps muscles, which are responsible for the negative lift required (see p. 61). In the final position the stabilization of the lumbar spine must be maintained, which amongst other things requires free flexion at the hip joints.

Horizontal Bending Type

If assessment of the relationships between lengths and the distribution of weights in a patient suggests that he may be of the horizontal bending type, we give the following instruction:
'Stand with your feet comfortably apart. Your bottom moves backwards, while your navel moves vertically downwards.'
Analysis: Because of the greater lower length and the bottom-heaviness of the upper length, the long axis of the body must be nearly horizontal for sufficient weight to be brought forwards. Because the thighs are long, their long axes are inclined backwards at only about 45°.
The weight of BSs pelvis, thorax and head (which are stabilized together) is distributed in the horizontal plane over the flexion/extension axes of the knee joints, so that little braking activity is required of the quadriceps, while the extensional stabilization of the vertebral column, particularly at lumbosacral junction, is considerable. However, since this is an automatic reaction, because only a stabilized long axis of the body can move enough weight forwards, there is no overstrain.

Notation

- Type (vertical or horizontal)
- Critical distance point, e. g. DP suprasternal notch, DP lower border of thoracic cage in the midfrontal plane, DP ear lobe, DP navel
- Inclination of the long axis of the body in the final position: horizontal/vertical or inclined $x°$ forward

> **Note**
> In bending down, the critical distance point is the point of the body which moves vertically downwards. It is determined in the final position, being the point which is situated over the part of the support area under most stress. Because this point has such an unambiguous, direct path, we can use it to give easily comprehensible instructions to the patient. It is situated on the upper length, between the iliac crest and the vertex, more cranial in vertical bending types and more caudal in horizontal types.

Fig. 308. No lumbosacral stabilization

Fig. 309. Thorax supported on the thigh

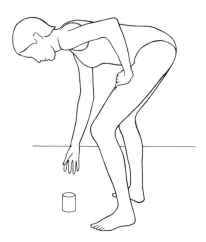

Fig. 310. A hand placed on the ipsilateral thigh or on a piece of furniture for support in bending

As the forward inclination of the body's long axis is recorded in degrees in our notation, we speak in terms not only of horizontal and vertical types but also of neutral types, which occur very frequently. As the long axis of the body inclines further forward, the stress on the lumbosacral region increases while that on the knee joints diminishes; as the long axis of the body returns to the vertical, the reverse happens.

Alternative Postures for Faulty Bending Behaviour

- Lunge position. This has the advantage of making backwards transfer of weight possible; it is advisable, therefore, to have the stronger leg behind.
- Alternatively, in the lunge position, the stronger leg can be brought forward and carry so much weight that the back leg goes into free play. As counterweight the latter provides activated passive buttressing, thus reducing the strain imposed upon the lumbar region. However, this entails some loss of the stability in this stance.

284

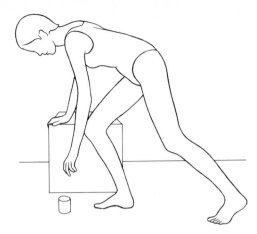

Fig. 311. Relief position in bending for a person with + upper length. The lunge posture with weight on the forward leg

Relief Postures in Bending Behaviour

- In standing with the long axis of the body inclined forward, the knees or the thighs can seek support at the work surface in front of them (see Fig. 137).
- In sitting, the abdomen can be leant against a supporting surface ('abdominal support', see Fig. 67 in *Therapeutic Exercises*).
- Harmful bending behaviour (Fig. 308).
- If, when working bent over, the thorax can be rested on the thighs, the strain on the lumbar region is completely eliminated. This posture, however, requires good mobility of the hip joints in flexion (Fig. 309).
- In bending down, a hand can seek support on the thigh of the same side or on a piece of furniture; this affords considerable relief from strain (Figs. 310, 311).

6.7 Respiration

In this section we consider *functional disorders of respiration* as a frequent result of insufficiency of postural statics. For discussion of functional respiratory disorders as a whole, the reader is referred to *Therapeutic Exercises*.

When the thoracic spine has lost its dynamic stabilization in the neutral position, it also becomes incapable of fulfilling its function of carrying the thoracic cage. In consequence, the normal costal movements of breathing are disturbed. The weight of the thorax hangs to a greater extent from the scaleni muscles and the positional relationship of the head to the thorax alters, the head being too far forward. This in turn causes additional changes of tone in the musculature. Reactive hypertonus arises in the cervical region. The abdomen is contracted to reduce the weight of the thorax attached to the scaleni. The consequence of the altered muscle activity is functionally defective breathing, in which hyperactivity of the scaleni is already present in normal breathing at rest. Costal breathing activity is reduced, because the scaleni raise the thoracic cage en bloc. The excursion of the diaphragm is also altered, since the inspirational expansion of the lower thoracic aperture cannot take place. During inspiration excessive swelling forward of the lower abdomen is visible.

6.8 Formulating the Functional Problem

The physiotherapist's findings during the assessment of a patient need to be brought into logical order. By interpreting the disorders of movement and pain as the result of deviations from hypothetical norms of constitution, mobility and postural statics, taking into account the medical diagnosis, we identify the *functional problem*. The formulation of the functional problem is the basis for a plan of treatment.

- The medical diagnosis sets quite specific limits to our therapy. The implication of these limits must be clearly expressed in the formulation of the functional problem, e. g. how much strain can be imposed on a fracture, a rheumatic joint, the respiration, the circulation, etc.
- Younger patients and children often have no pain or difficulties even when there are considerable deviations in the postural statics. It therefore takes very skilful tailoring of treatment to motivate them.
- In the anamnesis of pain one should differentiate between pain caused by the condition indicated in the diagnosis (e. g. progressive chronic polyarthritis) and pain that has its origin in the constitution and statics.
- Constitutional extra weights contribute to the functional problem if they are situated above the area of pain. They make themselves felt as soon as the long axis of the body moves out of the vertical, but they also need to be considered in a position of rest. For instance, if a patient in whom the frontotransverse diameter of thorax is relatively wide and the greater trochanters relatively close together has trouble in the lumbar region when lying on his side, he needs to have his pelvis and legs elevated.
- Deviations in the statics may even cause pain in the so-called erect posture. There are two basic differences to be made here:

  1. Shearing stress is imposed upon the passive structures when axial loading is no longer achieved at the joints. Ligaments and capsules are passive structures designed to react to proprioceptive impulses. To persistently strain them is to overstrain them; the reaction to this is pain with radicular symptoms, which enters the awareness of the patient. Pain is a warning signal, and if it is disregarded or suppressed with the help of medication, the protective structures, i. e. the discs and the cartilaginous tissues, are exposed to wear and tear, as occurs in neural compression or erosion of articulating surfaces. Pure shearing stress occurs when two weights deviate from the normal statics in opposite directions. The uppermost weight has no compensating counterweight above it, so its deviation brings about a different reaction in response to the disturbed statics of posture.

2. Reactive hypertonus is the normal reaction of sound musculature to poor posture and should be understood as braking or fall-preventing activity. Persistent hypertonus causes cramp, myogelosis and ischaemia, leading to ischaemic pain. Shearing stress and reactive hypertonus usually occur together.

Note
Static displacement of weights can cut out muscle activity as well as evoking it. The musculature that is 'switched off' is below and on the side towards which the weight has been displaced.
Reactive muscle activity works against the direction of the displacement of weight.

- Deviations from normal statics below the area of pain are involved in causing the pain, in that they make for a poor substructure. This is particularly so with deviations in the loading of the long axes of the legs, which make the legs a poor substructure for the vertebral column. Fixed poor position of the pelvis gives a poor substructure for the thorax, and a poor position of the thoracic spine gives a poor substructure for the shoulder girdle and the neck and head regions.
- Hypermobility is often the cause of poor statics, causing the vertebral column to develop areas of stiffness. However, these latter can also be the primary cause of deviations in the statics. We have to determine whether and how quickly they can be rectified.

6.9 Guidelines to Planning Therapy

Once the functional problem has been formulated, the guidelines for therapy are self-evident:

- Constitutional deviations from the norm cannot be altered. They have to be accepted, and it is up to the therapist to find appropriate relief postures. In cases where this proves impossible, the patient should consider changing his workplace.
- If conditional weights contribute to the constitutional problems, it is important to normalize them. Conditional surplus weights caused by well-exercised musculature have no adverse effect as long as they are compensated by an increase in strength and skill.
- Partial stiffness should be approached with great care. Success comes very slowly, and in the meantime we have to resort to relief postures.
- Hypermobility which cannot be controlled by muscular activity should be taken care of by orthopaedic aids.
- To influence posture, i.e. improve postural statics, the patient must be able to perceive the poor posture and the corrected posture. He must be able to feel not only when he has achieved the desired posture, but also when he is slipping back into the undesirable, habitual posture. As long as there is no stiffness, this can be achieved quite quickly. Correcting the posture cancels out the shearing stress and reactive hy-

287

pertonus of the musculature but is accompanied by hyperactivity, which brings about faulty respiration. The hyperactivity must be reduced after the correct posture has been taken up; for a successful result, the correct posture must be maintained despite reduction of the hyperactivity and until normal resting breathing begins. Correction of posture in standing is carried out while maintaining equal distribution of weight over the forefoot and the hindfoot. As the head is already correctly placed over the feet, the postural changes have to be brought about by displacement of weights at the joints of the hips and lumbar spine, and at the joints of the thoracic and cervical spine.

The therapist initiates only one displacement of weight; this automatically sets a counterweight in motion, because the distribution of weight over the feet must not be altered. Posture can be corrected starting from the head if the patient is sitting down. The critical distance point which will initiate the change and the distance point of the counterbalancing reaction must be identified, and whichever one will make it easiest for the patient to achieve the correct posture should be chosen for the movement instruction.

- Instructions in the use of relief postures for all situations in daily life are of paramount importance in every course of therapy and should be included at the patient's first consultation. Relief postures can be adopted even at work. It is important to employ such postures as soon as discomfort occurs. The answer to the patient's question of how often the relief posture should be adopted is simple: 'Whenever discomfort reminds you of the pain.'

6.10 Examples

The following method of notation of functional status has proved itself in practice:

Patient: Name and other personal details, referring physician, date of assessment

Functional Status: Name of physiotherapist

Condition: Diagnosis, purpose of referral, medical history

Constitution
    Height, weight, bone structure
    Lengths, widths, depths, distribution of weights

Mobility
    Vertebral column
    Other joints
    Lumbar kyphosis: Long sitting/stride sitting/cross-legged sitting
    Ischiocrural shortening
    Cervical kyphosis
    Shortening of pectoral or triceps muscles

Postural Statics
   From the side
   From behind/in front
      Distribution of weight in relation to the midfrontal plane/plane of symmetry
      Shearing stress on the passive structures of the joints
      Reactive hypertonus and hypotonus of the musculature

Leg Axes

Basic Gait Test

Bending Behaviour

Respiration

Functional Problem

Therapy

Case Histories

The examples below are set out as the physiotherapist would write them down during the assessment. A copy should be sent to the referring physician with compliments and thanks for the referral, and in this all abbreviations and symbols should naturally be explained or written out in full.

Patient S. F., 1945, ♂

Functional Status: S. K. V., 31. 3. 83

Condition
   D: Scheuermann's disease of middle TS; L5 slipping down onto S1 2 mm ventrally
   PoR: Establish functional status
   H: Patient has suffered minor back trouble for 15 years. In the last 3 years he has had less exercise for occupational reasons; he sits a great deal and does less sport.
   There is deep-seated lumbar pain on both sides radiating into the iliosacral joints, occurring mainly at night.
   Occupation: Carpenter. *Sport:* Skiing, walking. *Hobby:* 0
   Recent therapy: 2 Years ago heat treatment, massage; currently under treatment again

Constitution
   178 cm/80 kg/bone structure + + weight or BSs legs, thorax and on the forearms
   − Length of feet

Mobility

> *Vertebral column:* LATFLEX LS, TS R concave X/ROT TS on both sides XXX
> *Hip joints:* EXT just reaches the neutral position on R, slight flexion contracture on L. IR/ER on both sides 5–0–35
> *Knee joints:* EXT barely reaches the neutral position
> *Lumbar kyphosis:* Long sitting +, cross-legged sitting +

Postural Statics

From the side:
> − Longitudinal arches of the feet
> + FLEX of the pelvis at the hip joints with compensating + LS
> − TS (neck kyphosis cranially extended to C5)
> + CS (middle, upper)

From behind/in front:
> + Divergence of the long axes of the feet with slightly elevated pelvis R

Thoracic scoliosis: From in front: L broad, TS R concave, slight funnel chest. From behind: R broad with high shoulder L

Distribution of weight in relation to midfrontal plane/plane of symmetry: Pelvis, abdomen, head too far forward relative to thorax = predominance of ventral weights

Muscle tone: Lumbar hypertonus, paravertebral on both sides, reactive to the weight of the head and + FLEX of the pelvis at the hip joints

Functional Problem

1. The increased flexion of the pelvis at the hip joints with L5 slipping onto S1 and lack of extension at the hip joints increases the pathological tendency to slip and causes chronic persistent hypertonus at the lumbosacral junction.
2. In walking, insufficient rotation at the hip joints increases the persistent flexion of the pelvis at the hip joints and prevents the roll-on phase from occurring normally.
3. Through use of a discreet limping mechanism when walking, the patient has completely blocked rotation in the TS, such that all rotational demands in both walking and activity of the arms impose strain on the LS.
4. To sum up the problem, the patient is what you might call 'muscle bound'.

Therapy

1. Priority is intensive buttressing and lift-free mobilization of the hip joints and thoracic spine in rotation.
2. Instruction should be given in relief postures when lying. This will probably only be possible after the hip joints have been mobilized.

Patient M. L., 1962, ♂

Functional Status: S. K. V., 30. 3. 83

Condition
>   *D:* Pain bilaterally in shoulders, forearms, hands
>   *PoR:* Physiotherapy
>   *H:* Cello student, currently studying in Basel.
>   *Pain in both wrists,* R more than L, particularly in pronation and supination, oc-
>   curring intermittently for about a week at a time. Periods of pain more frequent
>   in cold weather in winter. Tests for rheumatic factors proved negative.
>   *Pain in neck and shoulder girdle,* L more than R, particularly after intensive prac-
>   tice or when heavy garments (thick pullover) worn when practising
>   *Pain in the small of the back* after long periods of practising and lack of sleep
>   *Occupation:* Cello student. *Hobby:* Drawing
>   *Recent therapy:* Chiropractic in the Netherlands for small of back and massage;
>   always produced short-term relief but no permanent improvement.

Constitution
>   192 cm/75 kg/bone structure normal
>   + Sagittotransverse $\emptyset$ at the level of the navel (flatulence)

Mobility
>   *Vertebral column:* Mobility relatively good. LATFLEX LS L concave X/TS R
>   concave X. Typical lordotic deformation of the vertebral column in the region
>   of T8, cranial sitting kyphosis
>   *Shoulder joints:* L FLEX/ABD/ER X
>   *Elbow joints:* Pronation/supination L X
>   *Lumbar kyphosis:* Long sitting + + to T9, cross-legged sitting and stride sitting
>   N. A. D.

Postural Statics
>   *From the side:*
>   Flat-back type
>   − Longitudinal arch of the feet
>   + FLEX at both hip joints
>   Long axis of BS thorax inclines sharply backwards, that of BS head forwards
>   *From behind/in front:*
>   Elevated pelvis R (almost disappears in sitting)
>   Slight scoliosis of the thorax due to ABD syndrome of L shoulder joint
>   *Distribution of weight in relation to midfrontal plane/plane of symmetry:* Pelvis, ab-
>   domen in front/thorax, shoulder girdle behind
>   *Muscle tone:* L lumbar hypertonus despite elevated pelvis R. Cervical hypertonus
>   more pronounced on L than on R, reactive to the weight of the arm. Slight atro-
>   phy of the musculature of the L leg

Leg Axes
>   + Tibial torsion with ER type hip joints and + valgus at the knee joints, L
>   more than R; elevated pelvis R only when standing.

Functional Problem
1. The poor statics of the patient's posture, combined with his constitutional tall-ness, causes reactive hypertonus in the cervical region even in the resting posi-tion. When the overstrain is increased by prolonged practice in a bad posture, pain not surprisingly occurs in the region of the strained musculature; this pain must be regarded as ischaemic.
2. The very poor sitting posture (increased extension of the pelvis at the hip joints and total flexion of the LS and TS, with compensating hyperextension of the CS) is partly due to the patient's height, which is reduced by this posture. The patient's posture ought to be analysed while he is practising.
3. The strongly developed musculature of the forearms with hypertonus of the overall musculature and the inability to relax sufficiently to achieve resting muscle tone causes (slightly) restricted movement in pronation and supination. As a consequence it becomes more difficult to set flexion/extension axes of the elbows and wrist joints parallel to each other, which explains the overstraining of the wrists during practice.

Therapy
Ice/mobilizing massage/instruction in relief postures/analysis of posture.

Patient W. N., 1963, ♀

Functional Status: C. M., 25. 10. 82

Condition
D: Lateral and medial epicondylitis and overstrain of musculature of L forearm; the patient is left-handed; she works as a shorthand typist
PoR: Treatment at the discretion of physiotherapist according to diagnosis. Pos-tural re-education, exercise programme
H: For the past 3 weeks, pain in L elbow radiating into the forearm, down to the wrist. Similar pain occurred 9 months ago, but was less pronounced.
Pain on pressure, particularly in the area over the lateral epicondyle, less inten-sive over the medial epicondyle. The pain increases when muscles are tensed and during movement. Patient also experiences pain in the thoracic and cervi-cal areas when sitting at work.
Occupation: Office clerk. *Sport:* Squash in winter. *Hobby:* Needlework
Therapy to date: Electrotherapy, no visible improvement; medication

Constitution
167 cm/72 kg/bone structure medium
+ Upper length (+ BS thorax)
+ Track width/+ distance between the greater trochanters/+ distance be-tween the shoulder joints

Postural Statics:
   From the side:
      Flat-back type with collapsed lower TS due to + anterior weight
      − Longitudinal arches of the feet with + plantar flexion at the talocrural joints; + FLEX of the pelvis at the hip joints and + + LS
      TS + lower, collapsed; − middle and upper
      CS: neck kyphosis to C6; from C5 + CS
   From behind/in front:
      High R shoulder with thorax slipped down R and slightly forward position of L shoulder
      Muscle tone: Overstrained rhomboid and levator scapulae muscles. 'Pincer jaws' of scapula and clavicle fit snugly on the thorax ·
      Hypertonus of the extensors at the wrist L gives pain

Leg Axes
      Tibial torsion absent with slight + varus of the lower legs

Arm Axes
      Pronounced + valgus of both elbows (approx. 15°)
      When placing FLEX/EXT axes of hand and elbow joints parallel with lower arms in pronation, L worse than R; hypermobility at the shoulder joints

Functional Problem
   1. The identifiable hypertonus, particularly of the extensors of the L wrist, enables us to define the problem as ischaemic pain.
   2. The poor postural habits in sitting, increased extension of the pelvis at the hip joints and total flexion of LS and TS, together with hypermobility of the shoulder girdle and the shoulder joints, have led to an over-long, unstabilized proximal 'lever for the forearm' of the working hand. This entirely explains the chronic overstrain of the musculature.

Therapy
   1. Intensive and repeated use of ice, followed by
   2. Intensive working of the hypertonal musculature; this skilful patient should be shown how to do this herself
   3. Dynamic stabilization of the TS in the neutral position
   4. PNF arm patterns with resistance proximal to the elbow
   5. Mobilizing massage of the shoulder girdle

Patient F. D., 1944, ♀

Functional Status: S. K. V., 17. 1. 83

Condition
   D: Asymmetry of the patella, Wiberg type II, III; pain due to overstrain, more on L than on R, and slight lateral dislocation of the patella

PoR: Training of the quadriceps and suitable exercise programme

H: Discomfort in L knee joint for several months. Piercing pain in the area of the lateral and medial space of the joint, particularly when sitting with the knees bent and when walking up- and downhill. Intermittent shooting pain, e. g. when getting on or off a tram or bus. Also arthritic changes in CS and LS, radiating into the shoulder joints, upper arms and the chest area.

Occupation: Housewife

Constitution

       180 cm/75 kg/bone structure normal

       + Upper length

Mobility

    Vertebral column: N. A. D.

    Hip joints: EXT just reaches the neutral position

Postural Statics

  From the side: N. A. D.

  From behind/in front:

       + Divergence of the long axes of the feet through lateral rotation of the head of tibia at the knee joint, with + eversion at the subtalar joints and slight medial rotation of the femoral condyles by external rotation in the knee joint

       Elevated pelvis R with slight scoliosis of LS/TS; bulge in the lumbar region L with − ROT of the lower thorax, slight asymmetry of the pelvis, R shoulder low and forward

  Muscle tone: Bilateral atrophy of the vastus medialis and the medial head of the triceps surae

Leg Axes

       + Valgus of R knee joint, + bilateral tibial torsion. Slight swelling of L lower leg

Functional Problem

    1. The poor statics of the axes of the legs, with increased bilateral tibial torsion and increased valgus of the R knee joint, combined with insufficient stability of the knees (external rotation of the lower leg is possible with the knee joint in the neutral position), has caused atrophy of the medial musculature of the calves and the thighs. It is significant that it is the stabilizing rotational components in the leg statics and musculature that are absent. This means that the knees, already weakened by the Wiberg asymmetry, are exposed to additional strain from the passive supporting structures. The patient's above-average height should not be forgotten as an important contributory factor in the malfunction.

    2. The asymmetry of the scoliotic position of the pelvis and vertebral column has somewhat reduced the patient's feeling for a sound pattern of movement in locomotion.

    3. The disturbances in the cervical region are adversely affected by the poor statics of the legs.

Therapy
1. Application of ice to the knees and cervical region
2. Lift-free training of quadriceps and flexors for the knee joint, first as buttressing activity in the lying position, then in sitting up with slight flexing of the knees by pressing down with the soles of the feet; there should at first be no leaning forward of the long axis of the body. Later, training during weight bearing by swaying backwards and forwards, with involvement of the talocrural joints
3. Gait training with a narrow track. Walking

Patient W. P., 1936, ♀

Functional Status: R. G., 10. 12. 82

Condition
D: Iliosacral joint syndrome, bilateral
PoR: Warm packs, twelve treatments
H: Patient has for 2 years had pain in the iliosacral region, particularly on the L, when the weather changes and during bending. She is a heavy smoker; speech and facial expressions slightly impeded (neurological disorder?)
Occupation: Waitress

Constitution
165 cm/64 kg/bone structure normal
+ Distance between the greater trochanters
+ Distance between the shoulder joints
Ventral weights predominate

Mobility
Vertebral column: Very good mobility overall
FLEX LS X/LATFLEX LS L concave X

Postural Statics
From the side:
Collapsed flat-back over slight + FLEX of the pelvis at the hip joints
From behind/in front:
Elevated pelvis R; patient stands back on L hindfoot

Basic Gait Test
Due to elevated pelvis R, weight is transferred asymmetrically from R to L and back again. When the R leg is weight-bearing, the R hip joint moves in adduction and the pelvis in backward rotation; when the L leg is weight-bearing, the pelvis moves in abduction

Functional Problem
1. The elevated pelvis R with predominant ventral weight, constitutional wide distance between the greater trochanters and collapsed flat back explain the disorder in the weight transferral and cause persistent overstraining of the iliosacral joints.

Therapy
  1. Warm packs
  2. Instruction in the use of relief postures for all situations and reduced-lift equilibrium exercises

Patient B. J., 1947, ♂

Functional Status: S. K. V., 5. 5. 83

Condition
  D: Lumbago
  PoR: Assessment of functional status, formulation of functional problem and suggested therapy
  H: Patient has suffered from back trouble since his youth. The first acute attack occurred in 1976 with the pain radiating into the loin R. Under the strain of weight bearing the pain spreads to the anterior aspect of the R thigh (L2). Pain on pressure over L5/S1, increasing in FLEX and L concave LATFLEX of LS. The R tensor and the attachments of the fascia lata of the abductors are painful on pressure, from ventral to the midfrontal plane, ischaemic in character.
  Occupation: Farmer and worker in a paper mill. *Sport:* Skiing
  Medication: Not effective

Constitution
  174 cm/63 kg/bone structure normal
  + Upper length (+ weight on thorax/head/arms)
  + + Frontotransverse ⌀ (slight funnel chest and thoracic scoliosis; narrower on L from behind and on R from in front with distinct inspiratory position of the ribs)

Mobility
  Vertebral column: FLEX L1–T10 X (above the sitting kyphosis) EXT L5–L2 X/ LATFLEX LS, L TS concave X/ROT TS positive XX
  Hip joints: IR/ER R 25–0–30, L 35–0–30
  Lumbar kyphosis: Long sitting: + extending to L2; cross-legged sitting: + +; stride sitting: N. A. D.

Postural Statics
  From the side:
    + FLEX of the pelvis at the hip joints
    LS (+ from L5 to L3)
    TS (+ up to T6, kyphosis extends caudally to L2, − up to C7)
    + CS
  From behind/in front: N. A. D.
    Distribution of weight in relation to midfrontal plane/plane of symmetry: The thorax is translated dorsally onto L2
    Shearing stress: + + Lumbar: the weight of the thorax is slipping backwards/ downwards from above onto L2

Muscle tone: Hypertonus of the striated abdominal musculature reactive to the weight of the thorax. There is slight lumbar hypertonus reactive to pain

Leg Axes
+ Tibial torsion, more on R than on L
+ Varus of L lower leg

Basic Gait Test
Walking on the spot: moderate persistent lumbar hypertonus and overloading of the tensor fasciae latae on the side of the supporting leg

Functional Problem
1. Due to the + weights on thorax/head/arms within the constitutional + upper length, overloading occurs in the lumbar region as soon as the long axis of the body inclines out of the vertical, which constantly happens in work on a farm or in a paper mill. Given the condition of the axes of the legs (+ tibial torsion, + varus of L lower leg), the patient should be advised to use the lunge position for bending down.
2. The poor statics of the vertebral column have imposed massive shearing stress upon the passive structures at the level of L1/L2; the pain radiating into the ventral aspect of the R thigh would be symptomatic of this.
3. The reactive disturbances of muscular tone – the persistent hypertonus of the striated ventral musculature, the lumbar musculature and the musculature which attaches the pelvis to the thighs – make potential mobility of the pelvis at the hip and lumbar joints impossible and cause ischaemic pain.

Therapy
1. Application of preparatory heat to LS/lower TS; perhaps ice treatment of the tensor fasciae latae.
2. Teaching of relief postures for all situations, to be used whenever discomfort is felt during the day.
3. Lift-free or reduced-lift mobilization of all joints where movement is restricted, followed by posture training to regain dynamic stabilization of the thoracic spine and potential mobility of the lordotic sections of the vertebral column and the hip joints.

Patient L. L., 1925, ♀

Functional Status: S. K. V., 22. 4. 83

Condition
D: Pain due to overstrain in the knee joints, more pronounced on R than L, with patella Wiberg type III–IV
PoR: Treatment at the discretion of the physiotherapist
H: Patient has had pain in the knees for about 2 years. More recently, swelling has occurred, more on R than L, but this reponded well to application of an ointment.
Occupation: Housewife

Constitution

167 cm/60 kg/ bone structure average to slight

+ Upper length (+ + BS thorax)

+ + Frontotransverse thoracic ∅ (+ weight on BS thorax)

Mobility

Vertebral column: Good mobility, tending to hypermobility and instability

Hip joints: IR/ER R 55–0–40, L 40–0–30

In 90° FLEX at the knee joints, pronounced ER of the lower legs at the knee joints

Lumbar kyphosis: Long sitting: + from S1 to T12 with R concave LATFLEX of the LS; cross-legged sitting: + +; stride sitting: N. A. D.

Postural Statics

From the side:

Flat-back type

− LS/ − − TS

From behind/in front:

+ Divergence of both long axes of the feet with + weight-bearing on R and standing back on R

The pelvis is slightly raised on R (even in sitting)

Muscle tone: Lumbar hypertonus reactive to the + weight on BS thorax, but no discomfort in the lumbar region

Leg Axes

+ Varus of R lower leg with − tibial torsion and increased detorsion at the hip joint. The varus/valgus line on R passes 1 cm lateral to the lateral malleolus. + Tibial torsion on L and slight detorsion.

Atrophy of the triceps surae and quadriceps muscles, particularly medially, more pronounced on R than L.

Basic Gait Test

1. In shoes with medium heels the roll-on of the weight-bearing foot is absent. Limping mechanism: premature medialization of weight bearing on the feet, causing flattening of the longitudinal arches in eversion
2. Without shoes, medial rotation of the FLEX/EXT axes at the knee joints occurs in the stance phase, particularly on the left

Functional Problem

1. The axes of the legs are asymmetrical due to differing tibial torsions and detorsion angles; the + varus of the R lower leg is pathological. The resulting disruption of the gait pattern by the premature medialization of weight bearing on the feet produces a substitute roll-on phase over the inversion/eversion axis. This limping mechanism overstretches the medial part and overcompresses the lateral part of the knee joint of the supporting leg.

Therapy

1. Preparatory application of ice
2. Instruction in the technique of ascending stairs, relieving the quadriceps by in-

creasing the forward inclination of the long axis of the body, which shortens the load arm of the extension at the knee joint

3. Quadriceps training, non-weight-bearing; functional weight-bearing exercise to strengthen the axes of the legs
4. Mobilization of lumbar kyphosis

Patient J. R., 1952, ♀

Functional Status: S. K. V., 20. 10. 82

Condition

D: Dorsalgia radiating more strongly L than R

PoR: Functional status, functional problem, suggestion for therapy

H: The patient experiences pain between the shoulder blades; extreme after birth of second child. Pain has been radiating into the legs since April 1982. The pain is most severe when the patient is sitting; it is always caused by weight-bearing strain. Pain cramp-like between the shoulder blades, more aching in the lumbar region.

Occupation: Housewife, two children, 6 and 4 years. *Sport:* Moderate. *Hobby:* Sews a great deal. *Medication:* None.

Therapy: Physiotherapy, about 30 sessions resulting in improvement; condition deteriorated after discontinuation of treatment.

Constitution

163 cm/55 kg/bone structure normal to slight

+ Upper length (+ BS thorax, pigeon chest)

Widths emphasized below the navel

Mobility

Vertebral column: FLEX of lower LS, T7–T3 X/EXT L3–T7 X/LATFLEX L3–T7 R concave XXX, L concave X/level of rotation far cranial (about the level of T6). Good predisposition for potential mobility of the vertebral column; acquired stiffness

Hip joints: EXT barely reaches the neutral position, more pronounced on L than R; internal rotation type

Lumbar kyphosis: Long sitting: + + extending as far as T7; stride sitting: +

Ischiocrural muscles: + Shortening, particularly of biceps femoris

Postural Statics

From the side:

− Longitudinal arches of the feet, more pronounced on R than L, with + plantar flexion at the talocrural joints, more pronounced on L than R; + EXT at the knee joints (genua recurvata) and spontaneous + FLEX of the pelvis at the hip joints.

The patient makes a conscious correction of the position of the pelvis, which results in + FLEX at the hip joints effected by both levers with marked development of the gluteal musculature

LS (+ lower S1, L5, L4, − upper)
TS (kyphosis extended caudally to L3 and cranially to C6 with − TS at segments T5, T6, T7)

From behind/in front:
+ Divergence of the long axis of the foot with frontotransverse position of the FLEX/EXT axis at the talocrural joint; medial rotation of the FLEX/EXT axes at the knee joints and + bilateral valgus. + Stress on L with slightly elevated pelvis L, negative ROT of pelvis and positive ROT of thorax; shoulders forward, more on R than L

Shearing stress:
In the genua recurvata, the whole upper weight is pushing backwards/downwards from above. Slightly lumbar: the weight of the thorax is pushing backwards/downwards from above.

Muscle tone: Bilateral lumbar hypertonus reactive to the weight of the head
Hypertonus of upper and middle trapezius reactive to the weight of the head

Leg Axes
Slight bilateral valgus deviation of the knee joints; slight tibial torsion L

Basic Gait Test
The limping mechanism, which is marked in barefoot walking, is insignificant when shoes with heels of about 5 cm are worn.

Functional Problem
1. The stiffness of the vertebral column between L3 and T6, probably acquired in connection with the pectus carinatum (pigeon chest), imposes caudal overstrain on the LS, iliosacral joints and the middle and upper TS. The cranial displacement of the level of rotation and the − TS at motion segments T7–T3 explain the cramp-like pain between the shoulder blades.
2. The poor loading of the axes of the legs with genua recurvata, a − longitudinal arch and the lateral divergence of the functional long axis of the foot, added to the reduced extension at the hip joints, the massive lumbar kyphosis and moderate shortening of the ischiocrural muscles, make the legs a poor substructure for the vertebral column.

Therapy
1. Stretching of the ischiocrural musculature and mobilization of the lumbar kyphosis, an exercise which the patient can practise by herself
2. Improving extension of hip joint, followed by correction of the genua recurvata
3. Functional training of quadriceps
4. Loading of the axes of the legs should always be practiced wearing shoes with heels
5. Reduced-lift mobilization of the vertebral column; attempt, using translation between the thorax and the pelvis to R and L, to bring the level of rotation down into the lower TS
6. Mobilizing massage at the level of movement shoulder girdle/thorax

7 Treatment Techniques

Treating restrictions of movement and painful areas of the body has given birth to methods which, while avoiding causing pain, still permit the musculature and tissues to be worked thoroughly, and bring about lasting improvement of the restricted joint mobility. These three techniques developed out of practical working with patients, but if we analyse them according to the criteria of functional kinetics, we see that they do follow the rules of economical motor behaviour and are based on the principles of continuing movement and the limitation of continuing movement by buttressing.

7.1 Mobilizing Massage

Definition: *Mobilizing massage* is a muscle massage. It is called 'mobilizing' because the musculature is not kneaded or pummelled with the joints in a particular position but is alternately stretched and relaxed as the joints are manipulated into different positions, and is worked at the same time.

For the patient, mobilizing massage is not a passive procedure but training in kinaesthetic and tactile perception.

Procedure

In mobilizing massage, we indicate to the patient the level of movement at which manipulation is to take place and the movement components or plane of movement in which the manipulation will be performed, and/or which muscle is to be stretched and relaxed. If we are dealing with a section of the vertebral column, we specify which; otherwise, the lever being moved is specified.

- *Examples*

Level of movement: R shoulder girdle/thorax (Fig. 312)
Moving lever: Shoulder girdle with right arm
Planes of movement: transverse, frontal, sagittal
Muscles: Upper trapezius and levator scapulae, lower trapezius and latissimus dorsi, middle trapezius, rhomboids, pectorals

Level of movement: Thoracic spine, in flexion and extensin (Fig. 313)
Muscles: Extensors of the thoracic spine

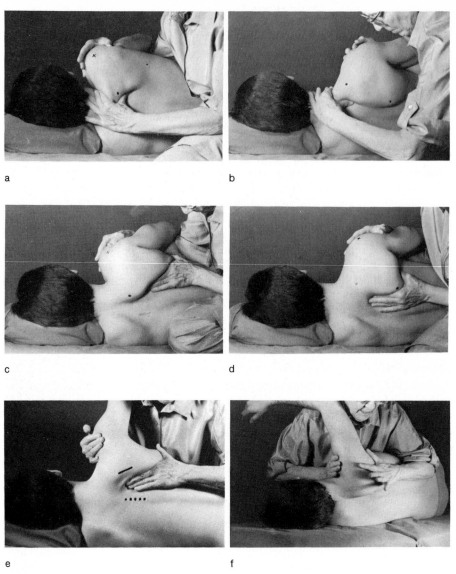

a

b

c

d

e

f

Fig. 312 a-i. Level of movement: shoulder girdle/thorax. **a,b** Upper trapezius and levator scapulae, **c,d** lower trapezius and latissimus dorsi, **e-i** medial trapezius, rhomboids, pectorals

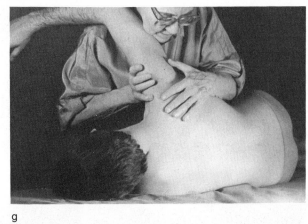

g

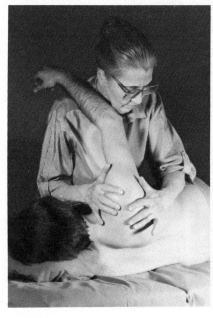

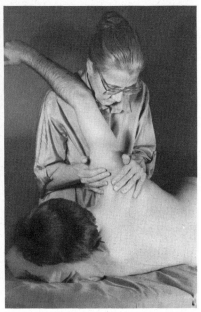

h   i

Fig. 312g–i. see page 302 for legend

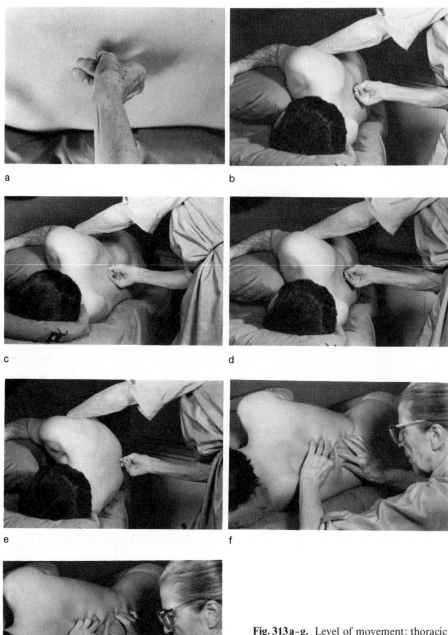

Fig. 313a–g. Level of movement: thoracic spine. **a** Position of the therapist's hand so as to exert axial pressure on the axis of movement, to cause extension of the thoracic spine; **b,c** yielding to the pressure; **d,e** pressure increased again by flexional activity of the patient in the thoracic spine; **f,g** mobilizing massage of extensors of the thoracic spine

Level of movement: Lumbar spine, in lateral flexion (Fig. 314)
Moving lever: Pelvis
Plane of movement: Midfrontal plane
Muscles: Oblique abdominal muscles, quadratus lumborum L

Level of movement: Thoracic spine/ribs, raising (inspirational) and lowering (expirational) (Figs. 315, 316)
Moving lever: Ribs
Muscles: Intercostals, levatores costarum, extensors, rotators, lateroflexors of the thoracic spine

Level of movement: Upper thoracic spine/cervical spine/atlanto-occipital and atlanto-axial joints, in flexion, extension, and ventral and dorsal translation (Fig. 317)
Moving lever: Head
Muscles: Paravertebral musculature of the upper thoracic and cervical spine and of the atlanto-occipital and atlanto-axial joints

Level of movement: Upper thoracic spine/cervical spine/atlanto-occipital and atlanto-axial joints, in flexion and extension (Fig. 318)
Moving lever: Head

Level of movement: Upper thoracic spine/lower cervical spine (Fig. 319)
Muscles: Upper trapezius, levator scapulae, paravertebral musculature in the region of C7

Level of movement: Upper thoracic spine/cervical spine, in translation to R/L; atlanto-occipital and atlanto-axial joints in lateroflexion (Fig. 320)
Moving lever: Head

Level of movement: Thorax/shoulder girdle/thoracic spine/cervical spine, in counter-rotational continuing lateroflexion (Fig. 321)

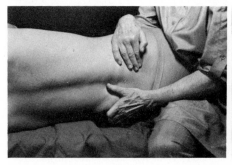 

a   b

Fig. 314a, b. Level of movement: lumbar spine. **a** Left concave lateroflexional manipulation, **b** right concave lateroflexional manipulation

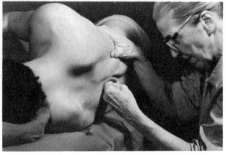

a

b

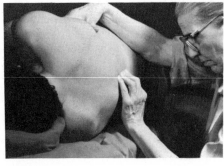

c

Fig. 315a-c. Level of movement: thoracic spine/ribs. **a** The therapist's left hand prevents lateroflexion as a continuing movement in the thoracic spine. **b** Lowering the rib cage (expiratory); **c** raising the rib cage (inspiratory)

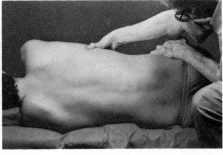

a

b

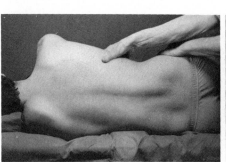

c

Fig. 316a-c. Level of movement: thoracic spine/ribs. **a** Starting position. The physiotherapist's hand is ready to draw the pelvis dorsally/medially. This causes a positive rotation of the pelvis in the thoracic spine, which allows the therapist's hand to raise the ribs at the thoracic spine by manipulation. As the ribs are lowered, the pelvis is rotated ventrally. **b** Inspiratory raising of the ribs; **c** expiratory lowering of the ribs

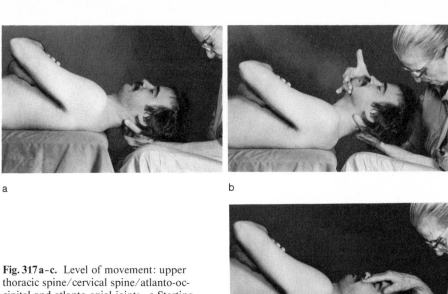

a

b

Fig. 317a–c. Level of movement: upper
thoracic spine/cervical spine/atlanto-oc-
cipital and atlanto-axial joints. a Starting
position. b Ventral translation, extension
at the atlanto-occipital and atlanto-axial
joints; c dorsal translation and flexion at
the atlanto-occipital and atlanto-axial
joints

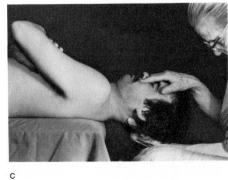

c

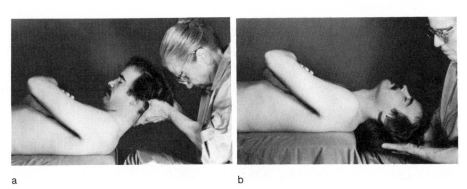

a

b

Fig. 318a, b. Level of movement: thoracic spine/cervical spine/atlanto-occipital and atlanto-axial
joints. a Flexion, b extension

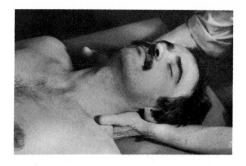

Fig. 319. Level of movement: upper thoracic spine/lower cervical spine. Position of therapist's hands for treating a neck kyphosis

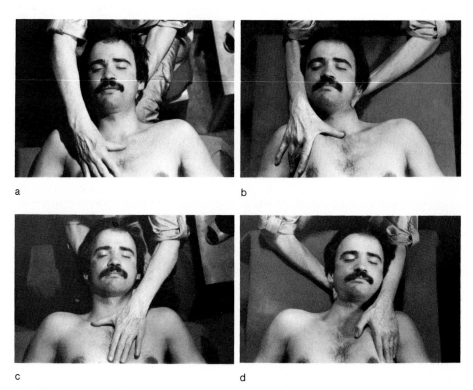

a   b

c   d

Fig. 320 a–d. Level of movement: upper thoracic spine/cervical spine/antlanto-occipital and atlanto-axial joints. **a** Starting position for the translation to the right; **b** translation to the right. **c** Starting position for the translation to the left; **d** translation to the left

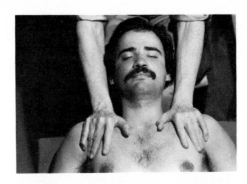

Fig. 321. Level of movement: thorax/ shoulder girdle. Starting position

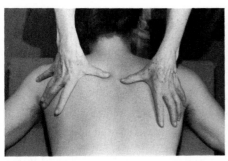
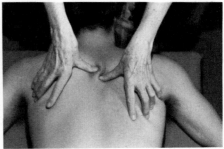

a

b

Fig. 322 a, b. Level of movement: upper thoracic spine/lower cervical spine. **a** Starting position. **b** Translation of the cranial spinous process to the right, and of the caudal spinous process to the left

Level of movement: Upper thoracic spine/lower cervical spine, in translation to R/L at the motion segments about C7 (Fig. 322)

Remark:
Figures 312-322 illustrate the positions of the physiotherapist's hands at the start and the end of manipulation, during which the musculature has been worked and relaxed. Figures 318-322 in particular show the method of mobilizing massage when treating a neck kyphosis.

Manipulating changes of position at the joints
- The joint or muscle which is to receive mobilizing massage is indicated to the patient.
- The physiotherapist determines the axis of the joint about which the mobilizing massage is to be performed, or the direction in which the muscle is to be stretched or relaxed, and decides which lever, pointer or gliding body is to be moved.
- The extent of the manipulation is defined by the available movement tolerance. This should not be completely exhausted, to prevent the occurrence of avoidance movements in the form of unwanted continuing movements.
- If it is necessary to use up the movement tolerance right up to end-stop, the technique of buttressing mobilization is required (see Sect. 7.2).

309

- Manipulation is end-stopped or buttressed partly by the weight of body segments not being moved, and partly by the physiotherapist, who must use manipulation to prevent an unwanted continuing movement.
- The musculature can be additionally worked by the physiotherapist during the relaxed phase.
- The physiotherapist must know exactly the origin, attachment, course and function of the muscles responsible for particular components of movement at the level of movement being worked. To determine the direction of the manipulation the course of each muscle concerned in the component of movement should be taken into account.

Note
There is no fixed starting position for either patient or physiotherapist in mobilizing massage of a level of movement, nor is any particular grip or position of the hands prescribed for the latter.

There are well-tried starting positions and methods of manipulation which are worth learning initially, but the size of the patient in relation to that of the physiotherapist offers so many potential variations that every physiotherapist must be able to make the necessary adjustments for every new patient. One point should always be remembered, however: the physiotherapist should not carry weight in her hands at the end of the long free levers of her arms, but should always seek support for her elbows against her own body, the patient or the treatment bench.

Training the patient's kinaesthetic and tactile perception
- Kinaesthetic perception training consists in raising the patient's awareness of distances and changes in distances between points on his body. Static kinaesthetic orientation locates regions of the body, e. g. 'the right hand', 'the back', 'the face', 'the sole of the left foot', and measures distances, e. g. 'shoulder to chin', 'left ear lobe to left iliac spine', 'point of elbow to navel'. Dynamic kinaesthetic orientation registers movements, e. g. 'the right shoulder is moving towards the right ear'; 'the knees are moving apart, but maintaining equal distance from the navel'; 'the tip of the nose is moving towards the right shoulder, which remains still, but the right hand touches the chin.'
- Tactile perception training is carried out through the body contact between the patient and the physiotherapist. Tactile perception tells the patient whether parts of his body are being pulled or pressed, whether the therapist has just started to touch him or just ceased, whether a pressure is increasing or decreasing. The physiotherapist's hands, in particular, provide the patient's body with a constant stream of changing tactile experiences.
- Under the verbal and tactile guidance of the physiotherapist, the patient is brought to perform or suppress precisely differentiated movements, and may thus succeed in spontaneously releasing a cramp which had previously seemed unconquerable.
- The tempo of the mobilizing massage should be very low initially. As physiotherapist and patient are in tactile contact, agreeing on a speed is not a problem.
The physiotherapist can feel when the patient has understood what the manipulation

is to be and is ready to let it take place. After a preliminary practice run, the tempo can be stepped up.

- Mobilizing massage can only function smoothly if the physiotherapist gives the patient the right information at the right time. Unambiguous instruction requires at least two items of information. Changing the distance between two points on the body, or moving a point on the body in a particular direction, requires instructions with just as many conditions as a therapeutic exercise if the manipulation is to be carried out with precision.
- Mobilizing massage should be carried out as nearly lift-free as possible (see p. 322) so that the musculature is not unintentionally activated against gravity, for lifting and weight bearing often cause pain and restrict movement, impeding fine movements.

Mobilizing massage can be carried out at any level of movement of the body and always proceeds in the manner described here.

7.2 Buttressing Mobilization

Definition: Mobilization of a joint, fulcrum, level of movement or switch point of movement is called 'buttressing' because it is built up on the observation criterion of the buttressing of a continuing movement. As it always concentrates on one single fulcrum, the buttressing must take place with in the fulcrum itself. In this way, movement tolerances can be exploited to end-stop. This type of mobilization should be performed as nearly lift-free as possible (see p. 322).
For the patient, buttressing mobilization of joints is not a passive procedure but training in kinaesthetic and tactile perception.

Procedure

In buttressing mobilization we indicate to the patient the joint at which mobilization is to take place, the components of movement, and the fulcrum or pivotal axis about which or plane of translation in which the mobilization is to be carried out.
How does buttressing mobilization of the joints take place? The mobilization is built up out of 'working units', a unit being the buttressing mobilization of a degree of freedom with two components of movement. Buttressing mobilization of a joint has at least as many units as the joint has degrees of freedom. We differentiate between buttressing mobilization of hinge, rotary, translatory and compression/distraction type joints.

- *Hinge type:* We identify the fulcrum, a distal distance point on the distal lever and a proximal distance point on the proximal lever. The most efficient buttressing mobilization is achieved when all three of these points are set in motion. If this can be done, either the two distance points can move away from each other with the fulcrum moving to a position between them, or the distance points can move towards each other with the fulcrum moving away from them, i. e. from the line between the distance points. If the levers which are to be moved are too long and/or too heavy or unwieldy, at least two points need to be moved. It is an advantage if one of the two points to be moved is the fulcrum. If the muscles involved span two or more joints, displacement of the fulcrum is a part of the normal motor behaviour.

- *Rotary type:* The axis of movement is the pivot and the levers are the proximal and distal pointers. The best form of buttressing mobilization in a rotary-type joint is counter-rotation of the two pointers. Another possibility, however, is to move just one pointer and prevent the other from moving. If this is the course chosen, the rotational axis should be moved so far (keeping always parallel to itself) that no additional, unwanted component of movement is mobilized at that level of movement.
- *Translation type:* The plane of movement is the gliding plane and the 'levers' are a proximal and a distal or a cranial and a caudal pointer (one on each gliding body). As in the rotary type, the two pointers move in opposite directions, or one moves against the other, which remains stationary.
- *Compression/distraction type:* The plane of movement is the plane of contact, the 'levers' are the load-carrying bodies of the contact surfaces. The contact surfaces can be drawn apart or pressed together, or the surface on one side of the contact can be held firm while the other is pulled away from it or pressed against it.

Lift-free buttressing mobilization is only possible if the physiotherapist assumes the weight of the part of the body moved. If unwanted muscular activity arises to overcome gravity, this is always an indication that the physiotherapist has not yet entirely 'got a grip on' the art of manipulating weights.

Because buttressing mobilization utilizes movement in contrary directions, movement tolerances can be exploited to end-stop. The key to success lies in reducing the need to work against gravity; in this way, an extension of previously restricted movement can be achieved astonishingly quickly. In properly performed buttressing mobilization of the joints, avoidance movements cannot occur.

In buttressing mobilization of hinge- and rotary-type joints, the proximal lever or pointer should always be set in motion first, since this has the smaller movement tolerance. It must be remembered, however, that the proximal lever must have another fulcrum with a parallel axis at its proximal end in order to be set in motion at all in the mobilization.

If reversible muscular contractures or pain are present, an attempt is made to improve the blood circulation and relax the musculature reactively through tension of the agonists in maximal shortening. Once end-stop has been reached by manipulation, the patient is asked to hold the position. As relaxation begins, the physiotherapist must be skilful in holding the weights as they become free, otherwise the fall-preventing activities will come into play at once, and the planned relaxation cannot take place. This relaxation, during which the weights are held by the physiotherapist, lasts about 30 s and passes through three clearly perceptible phases.

Training the Patient's Kinaesthetic and Tactile Perception: see p. 310

● *Example*
The seven working units of buttressing mobilization of the shoulder joint

Indications
- Restriction of movement at the humeroscapular joint
- Avoidance movements at the acromioclavicular and sternoclavicular joints and the motion segments of the vertebral column, particularly in the thoracic and cervical regions
- Periosteal pain at the points of origin of the musculature of the shoulder joints
- Pain of the tendon of the long head of biceps at the intertubercular synovial membrane. It should be remembered that the biceps does not normally have its point of origin at the humerus and the stress imposed upon it by the head of the humerus is quite considerable.
- Pain due to contraction and distraction of the musculature of the shoulder joint and the shoulder girdle
- Hyperactivity, shortening and reactive hypertonus of the musculature of the shoulder joint and the shoulder girdle (e. g. in cervical kyphosis)

Goals of Treatment
- Mobilization of the humeroscapular joints and acromio- and sternoclavicular joints
- Training the patient's perception of the functionally important buttressing function of the activity of the musculature of the shoulder joints and shoulder girdle when the hands are engaged in movements requiring skill
- Reducing the hyperactivity, which causes avoidance movements and overstrain of the cervical and thoracic spine
- Improving blood circulation in the periarticular structures
- Promoting diffusion in the cartilaginous tissues by 'cartilage massage'

Theory
The primary fulcra are the humeroscapular and sternoclavicular joints (there is concomitant movement at the acromioclavicular joint).
The distal moving lever is the humerus, the proximal moving levers are the scapula and the clavicle.
Possible rotary pointers are: distally, the flexion/extension axis of the elbow joint or, with 90° flexion at the elbow joint, the long axis of the forearm; proximally, the medial border of the scapula.
The movements are of the hinge and rotary type. Basically there are four variants:

1. Both levers perform a hinge-type movement (frontal abduction/adduction and transverse flexion/extension).
2. Both pointers perform a rotary-type movement, with the long axis of the upper arm frontotransverse (external/internal rotation).
3. The distal pointer performs a rotary-type movement, the proximal lever a hinge-type movement, with the long axis of the upper arm frontosagittal or sagittotransverse (external/internal rotation).
4. The distal lever performs a hinge-type movement and the proximal pointer a rotary-type movement (sagittal flexion/extension).

If hinge-type movements are brought about at both fulcra, the proximal lever and thus also the fulcrum at the shoulder joint can glide a long way on the thorax. If rotary-type movements are brought about at both fulcra, the proximal pointer and thus the ful-

313

crum at the shoulder joint can only glide a little way on the thorax. If a mixture of rotary-type and hinge-type movements is brought about at both fulcra, the extent to which the shoulder joint glides on the thorax depends on the type of movement performed by the proximal lever/pointer.

In the combination 'proximal hinge-type/distal rotary-type movement', which implies considerable gliding of the fulcrum of the shoulder joint on the thorax, the axis of rotation (long axis of the upper arm) must move with the fulcrum, keeping parallel to its position at the start.

Technical hints

There are no prescribed holds for this technique. However, one hand must manipulate the scapula; the physiotherapist should try to control the gliding movement of the scapula on the thorax from the lower angle of the shoulder blade or from the acromion, by fixing her fingers on the spina scapulae.

In assuming the weight of the arm, it is often easier if the elbow is flexed to approximately 90°. However, this often gives rise to avoidance movements, since the weight of the forearm works in favour of internal rotation and/or extension at the humeroscapular joint.

It is not absolutely necessary but in most cases advisable that the patient lie on his side, resting on the unaffected shoulder. Head and waist should be well supported to enable BSs pelvis, thorax and head to be aligned in the long axis of the body. With the patient in this position the physiotherapist should preferably stand or sit behind him, except when the long axis of the upper arm is sagittotransverse, when it is better to stand in front of him.

If the mobility of the humeroscapular joint is greatly restricted, a starting position for this joint must be selected which will allow at least the proximal lever/pointer to utilize almost its full movement tolerance. As this movement is in any case the first to be manipulated, any remaining movement tolerance can be exploited by buttressing from the distal lever.

Practice

First of all, the humerus and shoulder girdle are moved to and fro about the desired axes in co-rotational continuing movement. The final position reached can be actively held by the patient.

Next, the humerus and shoulder girdle are again brought to an end-stop position by co-rotational continuing movement. The scapula is manually fixed in this position and the humerus is manipulated in the opposite direction in a buttressing movement. The same is done in the opposite direction. This final position can also be actively held by the patient, resulting in activation of the muscle in a position of maximal contraction.

Only after these introductory movements can the real buttressing mobilization start. The levers/pointers are moved in opposite directions, starting with the proximal lever/pointer, which has the smaller movement tolerance. The spatial path of dDP olecranon is now much shorter because the movement tolerance of both levers/pointers is being utilized.

The speed of movement is always slow at first. One aims at normal walking speed (120/min), making about 10-15 to-and-fro movements per working unit.

Working Units of Buttressing Mobilization

Unit 1: Abduction/adduction at the shoulder joint (Fig. 323)

a) Hinge-type movements at the sternoclavicular and humeroscapular joints with the levers moving in the plane of the scapula (approximately frontal) about sagittotransverse axes.

b) The distal distance point of the sternoclavicular joint, dDP acromion, moves cranially/caudally and medially/laterally, drawing the lower angle of the shoulder blade away from the thoracic spine and back to it.

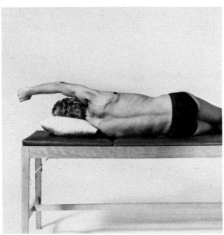

a

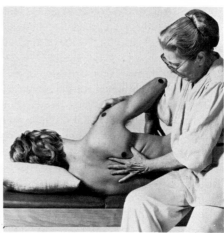

b

c

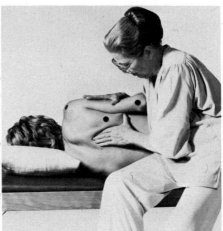

d

Fig. 323a–d. Unit 1: Abduction/adduction at the shoulder joint. **a** Abduction at the right shoulder joint, continuing movement at the shoulder girdle; **b** Buttressing mobilization in abduction. **c** Adduction at the shoulder joint, continuing movement at the shoulder girdle; **d** buttressing mobilization in adduction

315

c) Simultaneously, the distal distance point of the humeroscapular joint, dDP olecranon, moves cranially/caudally and medially/laterally, bringing about abduction/adduction at the shoulder joint.

Unit 2: Internal/external rotation at the shoulder joint (Fig. 324)
a) Hinge-type movement at the sternoclavicular joint and rotary-type movement at the humeroscapular joint with the proximal lever and the distal pointer moving in frontal planes about sagittotransverse axes.
b) As 1 b)
c) Simultaneously, with the long axis of the upper arm sagittotransverse, the distal distance point of the humeroscapular joint, dDP olecranon, moves cranially/caudally and medially/laterally, bringing about internal/external rotation at the shoulder joint.

Unit 3: Transverse flexion/extension at the shoulder joint (Fig. 325)
a) Hinge-type movement at the sternoclavicular and humeroscapular joints with the levers moving in transverse planes about frontosagittal axes.
b) Simultaneously, the distal distance point of the sternoclavicular joint, dDP acromion, moves medially/laterally and ventrally/dorsally, drawing the medial border of the shoulder blade away from the thoracic spine and back towards it.
c) Simultaneously, with the long axis of the upper arm frontotransverse, the distal distance point of the humeroscapular joint, dDP olecranon, moves medially/laterally and ventrally/dorsally, bringing about transverse flexion/extension at the shoulder joint.

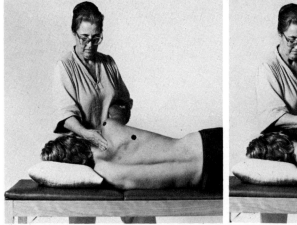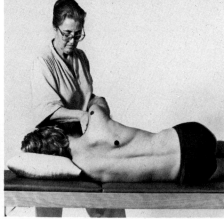

a   b

Fig. 324a, b. Unit 2: Internal/external rotation at the shoulder joint. **a** Buttressing mobilization in external rotation with the long axis of the upper arm sagittotransverse. **b** Buttressing mobilization in internal rotation with the long axis of the upper arm sagittotransverse

316

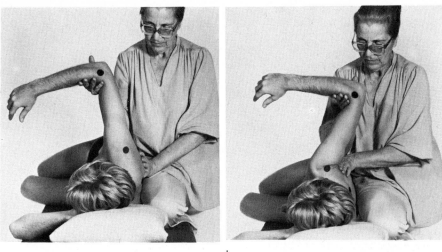

a   b

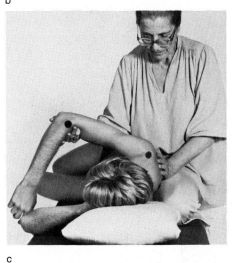

Fig. 325a-c. Unit 3: Transverse flexion/extension at the shoulder joint. **a** Starting position. **b** Buttressing mobilization in transverse extension. **c** Buttressing mobilization in transverse flexion

c

Unit 4: internal/external rotation at the shoulder joint (Fig. 326)

a) Hinge-type movement at the sternoclavicular joint and rotary-type movement at the humeroscapular joint with the proximal lever and the distal pointer moving in transverse planes about frontosagittal axes.

b) As 3 b)

c) Simultaneously, the distal distance point of the humeroscapular joint, dDP olecranon, moves medially/laterally and ventrally/dorsally, bringing about internal/external rotation at the shoulder joint.

Unit 5: Internal/external rotation at the shoulder joint (Fig. 327)

a) Rotary-type movement at the sternoclavicular and humeroscapular joints with the pointers moving in sagittal planes about frontotransverse axes.

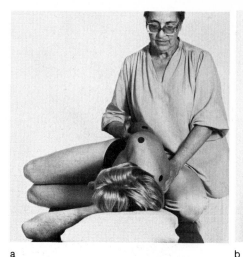

a

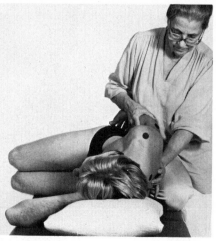

b

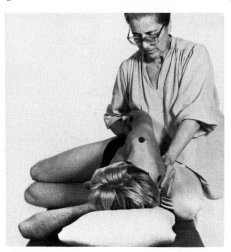

c

Fig. 326a–c. Unit 4: Internal/external rotation at the shoulder joint. **a** Starting position. **b** Buttressing mobilization in external rotation with the long axis of the upper arm frontosagittal. **c** Buttressing mobilization in internal rotation with the long axis of the upper arm frontosagittal

b) Simultaneously, the distal distance point of the sternoclavicular joint, dDP acromion, moves ventrally/dorsally and cranially/caudally, drawing the lower angle of the shoulder blade away from the thoracic spine and back to it.

c) Simultaneously, with the long axis of the upper arm frontotransverse, the distal distance point of the humeroscapular joint, dDP olecranon, moves ventrally/dorsally and cranially/caudally, bringing about internal/external rotation at the shoulder joint.

Unit 6: Flexion/extension at the shoulder joint (Fig. 328)

a) Rotary-type movement at the sternoclavicular joint and hinge-type movement at the humeroscapular joint, with the proximal pointer and distal lever moving in sagittal planes about frontotransverse axes.

318

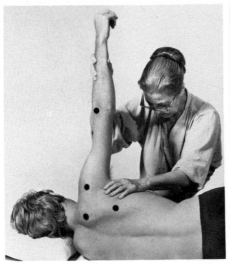

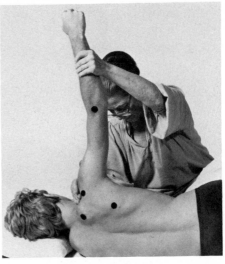

a   b

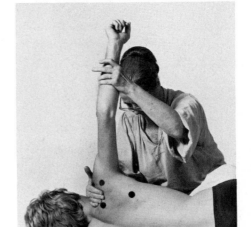

Fig. 327 a–c. Unit 5: Internal/external rotation at the shoulder joint. **a** Starting position. **b** Buttressing mobilization in external rotation with the long axis of the upper arm frontotransverse. **c** Buttressing mobilization in internal rotation with the long axis of the upper arm frontotransverse

c

b) As 5 b)
c) Simultaneously, the distal distance point of the humeroscapular joint, dDP olecranon, moves ventrally/dorsally and cranially/caudally, bringing about flexion/extension at the shoulder joint.

Unit 7: Flexion/extension at the elbow joint (Fig. 329)
As the biceps does not normally have its point of origin at the humerus but plays an important role in disturbances of movement at the shoulder joint, buttressing mobilization of the shoulder joint should not left unconsidered.

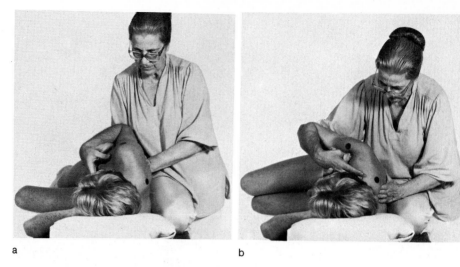

a

b

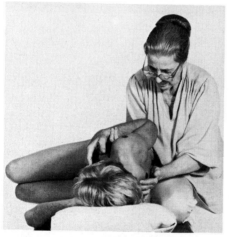

c

Fig. 328 a–c. Unit 6: Flexion/extension at the shoulder joint. **a** Starting position. **b** Buttressing mobilization in flexion. **c** Buttressing mobilization in extension

- The elbow joint is the primary fulcrum; movements involved in buttressing mobilization are pronation/supination of the forearm and movements of the humeroscapular and sternoclavicular joints in all degrees of freedom.
- The moving levers are, proximally, the forearm and, distally, the upper arm, scapula and clavicle.
- The movements taking place are of the hinge and rotary type.

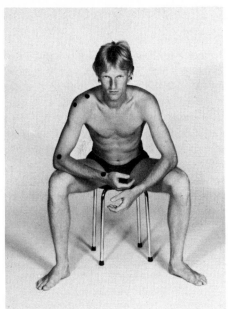

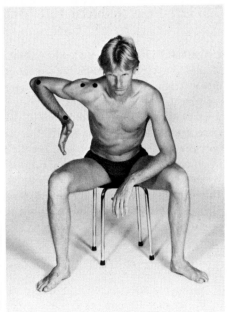

a

b

Fig. 329 a–c. Unit 7: Flexion/extension at
the elbow joint. **a** Starting position. **b** Buttressing mobilization in flexion. **c** Buttressing mobilization in extension

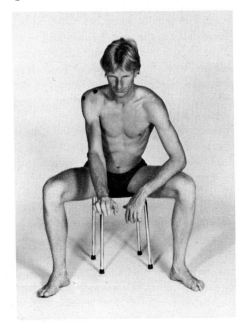

c

7.3 Lift-Free/Reduced-Lift Mobilization of the Vertebral Column

Definition: In lift-free/reduced-lift mobilization of a level of movement, we think first of all of the weight of the body segment or segments which are to be moved. When one lifts a weight, one is lifting it against gravity. We use the word 'lift-free' to describe movements in which parts of the body are moved without having to lift their own weight against gravity, or 'reduced-lift' if only a small fraction of the weight has to be lifted against gravity. If the lift-free movement of body segments is across a base support, the physiotherapist must ensure that the resistance due to friction is insignificant. In principle, lift-free mobilization is possible for every joint.

When selective treatment is required, lift-free mobilization is limited by buttressing. This technique is particularly suitable for treatment of the area of the vertebral column. The transition from being moved by the therapist to independent active movement can be effected smoothly (see *Therapeutic Exercises, Sect. 5.1*).

Procedure

In lift-free mobilization the axis of movement is vertical. It follows that lift-free mobilization of the vertebral column in flexion/extension takes place with the patient lying on his side, in lateroflexion with the patient lying supine (or, occasionally, prone), and in rotation with the long axis of the body vertical.

Translations to the right/left are lift-free in the supine position and when the long axis of the body is vertical; ventral/dorsal translations are lift-free when the body is on its side and when the long axis of the body is vertical.

Frictional resistance to translations in the supine or side-lying positions tends to be quite high, but this can be solved by placing a hand-towel under the body segment concerned.

Rotations and translations with the long axis of the body vertical have no frictional resistance to overcome. However, the body segments aligned in the long axis of the body need to be supported to prevent them from falling.

In lift-free mobilization of the vertebral column either the pelvis is moved at the joints of the hips and lumbar spine or the thorax is moved at the joints of the thoracic and cervical spine. Movements of the pelvis are buttressed in the thoracic spine, those of the thorax in the lumbar spine and at the atlanto-occipital and atlanto-axial joints.

If all the movement components are mobilized from the same starting position, some of the mobilization will be lift-free, some only reduced-lift. The physiotherapist can decide how she would like to proceed in such cases.

In *lying on the right/left side,* mobilization in flexion/extension and ventral/dorsal translation is lift-free and mobilization in lateroflexion and rotation is reduced-lift. Translation to the right/left is contraindicated because it demands too much lift.

In supine lying (or, more rarely, *prone lying*), mobilization in lateroflexion and translation to the right/left is lift-free, mobilization in flexion/extension and rotation reduced-lift. Ventral/dorsal translation is contraindicated because it demands too much lift.

When the long axis of the body is vertical (BSs pelvis, thorax and head aligned in this axis), mobilization in rotation and translation forwards/backwards and right/left is lift-free, and mobilization in flexion/extension and rotation is reduced-lift.

When the patient has been placed in the correct starting position, the physiotherapist tells him which are the critical distance points for the mobilization and in what direction they are going to move, she indicates a fixed point of reference on the body, and she gives the *conditio* for the limitation of the movement through active buttressing.

- *Example*

Aim: lift-free mobilization of the cervical spine through translation to the right/left.

Starting position: Lying supine, BSs pelvis, thorax and head aligned in the long axis of the body. BS head should if possible be in the neutral position, parked; BS legs is parked with slight flexion at the knee and hip joints supported by pillows. The tips of fingers II–V rest on the long axis of the sternum.

The physiotherapist grips the thorax on the right and left side in both hands, slightly cranial to the frontotransverse diameter of thorax, and asks the patient to feel through his fingertips how his sternum is being pushed in a gliding translatory movement about 1 cm to the right. His head does not move; neither do his pelvis and legs. He is returned to the starting position and then, without pausing, the same gliding movement is performed to the left, and so on (see *Therapeutic Exercises,* Sect. 5.1).

As soon as the physiotherapist feels that the patient is able to continue the movement without her guidance, she removes her hands. Unaided, the patient must sense the movement before the lift-free or reduced-lift mobilization commences by touching the distance points and calling to mind the exact direction in which they are to move. This brief training in perception paves the way for the movement, cutting out unwanted braking activities.

8 The Concept of Movement Training

The essence of functional kinetics is simple. As we said at the beginning, our model is the normal motor behaviour of the healthy human being. All our endeavours to use movement therapeutically end in the recognition that learning to approach the normal and healthy is like training in economy. This type of economy has nothing at all to do with penny-pinching or holding back on energy. On the contrary: in working with a patient, we are constantly seeking the limits of the possible: that is where practice starts to be effective, and if it is successful, the limits of the possible are extended still further.

First, we find a pain-free starting position for the patient. When we analyse this position, we find that the painful areas have been relieved of stress. Then we treat the patient. During treatment we converse with him; he observes what is being done and tells us how he experiences the treatment. Gradually the physiotherapist carries out the treatment less with her hands than with her words. Finally the patient is no longer a patient; the therapy is complete. What has he learnt? No more than to use the gifts which nature has given him.

The concept of movement training which developed from this model can be defined as economy through minimized lifting and dynamic stabilization.

This concept is developed in the practical books we have already mentioned, which are currently available in German (see References). An English-language edition of *Therapeutic Exercises* is in preparation, and translations of the other two books are planned.

9 Glossary

Actio
: In a movement sequence, the *actio* is the primary movement; it leads to the movement goal and dominates the patient's awareness. The *actio* displaces weights in the direction of movement; if it includes horizontal components, it has an accelerating effect on the movement.

Active insufficiency
: A muscle is actively insufficient if it is unable to contract sufficiently to actively fix, in the end-stopped position, the levers, pointers or gliding bodies of the switchpoint it bridges.

Activity state
: The multiplicity of possible postures and movements and their position in space under the influence of gravity demands different states of activity of the musculature. When we have defined these various activity states, we are in a position when analysing posture and movement to identify them, relate them to particular body segments, name them, and bring them about.

Angle of the body diagonals
: The caudal or cranial angle formed where the body diagonals cross.

Antetorsion, angle of
: If you lay a femur on a table, its neck points inwards/upwards. The angle formed between the axis of the femoral neck and the axis of the femoral condyles is the angle of antetorsion.

Avoidance mechanisms, avoidance movements
: Uneconomical, unwanted continuing movements deviating out of the direction of movement, changes in the support area, or buttressing of continuing movements.

Axis of movement
: The place at which levers and pointers rotate, i.e. the place at which movements of the joints take place.

Base plane
: Transverse plane tangential to the soles of the feet.

Basic gait test
: Test which may be carried out during the assessment of the functional status in order to analyse limping mechanisms.

Bridging activity
: When, in any body posture or movement, the support area is determined by more than one point of contact between body and base support, the body segments or parts of them which provide the contact with the base supports form bridges with their neighbouring body

	segments. The muscle activity which braces the arch of this bridge is called bridging activity.
Body diagonal	Line connecting the midpoint of one hip with the midpoint of the contralateral shoulder.
Body segment	Each functional body segment has several levels of movement, whose motor behaviour may be regarded as a functional unit.
Buttressing mobilization	Buttressing mobilization of a joint always concentrates on a single fulcrum. The buttressing must take place in the fulcrum itself. In this way it becomes possible to exhaust the movement tolerance up to end-stop. This mobilization should if possible be performed lift-free, or at least with reduced lift.
Buttressing of continuing movement	Limitation of a continuing movement by counterweighting, counteractivity or countermovement.
Buttressing of continuing movement, activated passive	Automatic engagement of muscle activity to adjust the lever arm(s) of the counterweight of a primary movement to the length needed. Buttressing a continuing movement with a counterweight is an automatic equilibrium reaction. The counterweight works against the horizontal component of the primary movement and has a braking effect upon it.
Buttressing of continuing movement, active	Buttressing of continuing movement by antagonistic muscle activity.
Buttressing of continuing movement, passive	Purely passive buttressing rarely occurs, because the parts of the body used for the buttressing are inherently mobile.
Caudal/caudally	'Caudal' indicates position and means 'on the feet or the part containing the feet'; 'caudally' indicates direction and means 'towards the feet or in the direction in which the feet point'.
Cervical kyphosis	Kyphosis in the area of the upper thoracic and lower cervical spine, which involves stiffness of these parts of the vertebral column and has an unfavourable effect on the postural statics of the head. Also called 'neck kyphosis'.
Concentric isotonic	Describes active contraction of a muscle.
Conditio	In a movement sequence, the *conditio* is the sum of the conditions which make define the movement finely and precisely. The patient must be conscious of the *conditio* and this will make it easier for him to perform movements with precision.
Condition	The influence which a patient's social position and psychic and somatic state have upon his motor behaviour.
Constant-location movement	Describes movement sequences in which either the

	support area does not change at all or changes take place only within the support area.
Constitution	The influence which the lengths, widths, depths and weights of a patient's body segments have upon his motor behaviour.
Continuing movement	When any given point on the body is moved by a movement impulse in a particular direction and movement excursions occur at neighbouring joints which help to bring about this movement, a continuing movement has taken place.
Cranial/cranially	'Cranial' indicates position and means 'on the head or the part containing the head'; 'cranially' indicates direction and means 'towards the head or in the direction in which the head points'.
Degree of freedom	One way in which a lever, pointer or gliding body can move to and fro at its joint connection. Each degree of freedom has two components of movement.
Detorsion	The normal process of reduction of the natal angle of antetorsion during development to adulthood.
Distal/proximal	These terms relate to the functional centrepoint of the body. 'Distal' means 'further from the body's centrepoint', proximal means 'closer to the body's centrepoint'.
Distance points	The distance points of a joint movement are the furthest points on levers and pointers from the axis of movement.
Eccentric isotonic	Describes active extension of a muscle.
Effectors	From the functional point of view, muscles are effectors of posture and movement. They can act as lifters and movers of weights, as brakes on falling weights, and as preventers of weights from falling.
End-stop, end-stopping	Limitation of joint mobility by the passive structures of the movement apparatus.
Equilibrium reaction	When displacement of the centre of gravity threatens a person's equilibrium, reactions using counterweighting, counteractivity, countermovement and/or a change in the support area ensure that the equilibrium is maintained.
Extensional activity	Activity of the muscles normally involved in extension at a joint.
Flexional activity	Activity of the muscles normally involved in flexion at a joint.
Free play function	Activity state which arises when an extremity is proximally suspended from the body and can move its distal end freely.
Frontal plane	Any number of parallel planes can be positioned between the front and back sides of the homuncu-

lus' cube. Where they cut the homunculus they divide him into a ventral and a dorsal part. All these planes are frontal. They relate to the body and not to space.

Frontosagittal axes
Lines of intersection of the frontal and sagittal planes of the body which pass through the centre of the joints of the vertebral column and proximal extremities.

Frontotransverse axes
Lines of intersection of the frontal and transverse planes of the body which pass through the centre of the joints of the vertebral column and proximal extremities.

Functional centrepoint of the body
Point of intersection of the two body diagonals.

Functional kinetics
Technique of direct observation and analysis of human posture and movement.

Hanging activity
Activity state which arises when the whole body or particular body segments or parts of segments hang from a suspension device; traction occurs at the joints involved.

Horizontal plane
A plane relating to space. When a person stands upright, his transverse planes are horizontal; on his back or his stomach, his frontal planes are horizontal; on his side, his sagittal planes are horizontal.

Isometric
Describes muscle activity which prevents possible movements: the length of the active muscle does not change.

Lift-free movement
Movement which does not require the parts of the body involved to be lifted against gravity.

Limitatio
In a movement sequence, the *limitatio* is the effect of the *conditio*.

Location-changing movement
Describes movement sequences in which the whole movement system of the inherently mobile human body moves to another place and has a new support area.

Long axis of the body
Line of intersection between the plane of symmetry and the midfrontal plane. It passes through the body's centrepoint and the vertex.

Lumbar kyphosis
Kyphotic alteration of the vertebral column at the lumbosacral junction. It may extend as far as the lower thoracic spine. A distinction is made between structural and functional lumbar kyphoses; the latter disappear when the ischiocrural musculature is relaxed. A lumbar kyphosis has a bad influence on posture and is often the cause of loss of potential mobility of the pelvis at the hip and lumbar vertebral joints.

Midfrontal plane
Frontal plane which passes through the body's centrepoint.

328

Midtransverse plane	Transverse plane which passes through the body's centrepoint.
Mobile	Body segment which, in a given posture or movement, is predominantly potentially mobile.
Mobility	The extent of passive and active movement tolerance at the joints.
Neck kyphosis	*See* Cervical kyphosis.
Observation criterion	A feature which has been isolated by systematic observation and manipulation of the human body at rest and in motion, and which helps to differentiate the pathological from the normal.
Observer's bisecting plane	This is the vertical plane of symmetry of the eyes of the observer, projected until it meets the patient, bisecting him into a right and a left part. If the observer aligns her bisecting plane with the patient's centre of gravity, she can assess the potential accelerating and braking weights of the patient's body in movement in a direction at right angles to the bisecting plane.
Observer's horizontal plane	This is the horizontal transverse plane through the eyes of the observer, projected forwards until it meets the patient, dividing him into an upper and a lower part.
Observer's parallel plane	This is the vertical frontal plane through the eyes of the observer, displaced forwards in a parallel movement until it meets the patient. It enables one to judge actual distances on the patient's body and make comparisons between them.
Observer's planes	The observer's planes help the therapist to avoid being misled by perspective in her optical perception of the patient. The observer's eyes must be horizontal and their plane of symmetry vertical.
Parking function	Activity state present in a body segment or part of it which is in contact with a base support and exerts on this only the pressure of its own weight.
Passive insufficiency	A muscle is passively insufficient if it cannot be stretched far enough to allow levers, pointers or gliding bodies at the level of movement it bridges to move until end-stopped.
Plane of symmetry	Sagittal plane through the body's centrepoint, also called the median or midsagittal plane.
Plane of the vertex	Transverse plane tangential to the vertex.
Potential mobility	Readiness of muscles to respond to stimulation with movement, i.e. with changes in the position of the joints.
Pressure activity	Muscle activity which increases the pressure exerted by a body at a point of contact with a base support.

329

	Pressure activity causes compression at the joints involved.
Primary movement	The initial movement towards the stated movement goal.
Proximal/distal	*See* Distal/proximal
Pushing off activity	Muscle activity which makes use of a point of contact between the body and a base support or supportive device to achieve a purposive thrust.
Reactio	In a movement sequence, the *reactio* is the name given to the automatic equilibrium reactions which arise in response to the primary movement or *actio*.
Reduced-lift movement	Movement in which as little as possible of the weights of the body is lifted against gravity.
Sagittal plane	Any number of parallel planes can be positioned between the left and right sides of the homunculus' cube. Where they cut the homunculus, they divide him into a right lateral and a left lateral part. All these planes are sagittal. They relate to the body and not to space.
Sagittotransverse axes	Lines of intersection of the sagittal and transverse planes of the body which pass through the centre of the joints of the vertebral column and proximal extremities.
Stabile	Body segment which, in a given posture or movement, is predominantly stabilized.
Stabilization	Muscular fixation of one or more joints in a given position.
Statics, postural	The influence which a patient's posture has on his motor apparatus with regard to stress.
Support area	The smallest area which includes all the points of contact between activated body segments and their base support.
Supported leaning activity	Activity state which, when the body is leaning against a supportive device, arises in the braced muscles on the side of the body facing the support.
Supporting function	Activity state which arises when one of the extremities is in contact with a base support and exerts more pressure on it than is due to its own weight.
Switchpoint of movement	This term emphasizes how, in functional kinetics, the interest of the joint is that it is the place at which movements occur through changes in the position of levers, pointers and gliding bodies.
Transverse plane	Any number of parallel planes may be positioned between the base plane and the plane of the vertex of the homunculus' cube, each dividing him into a cranial and a caudal part. All these planes are transverse. They relate to the body and not to space.

Vertex	Point of intersection of the plane of symmetry, the midfrontal plane and the plane of the vertex.
Virtual body axes	These are axes which exist only in the imagination but which, within the inherently mobile system of the body, are brought into being by maintenance of a particular posture. Thus the long axis of the body is formed by a particular arrangement of the pelvis, thorax and head, or the long axis of the foot is that (imaginary) one about the foot arcs when twisting from resting on the medial to resting on the lateral border. By contrast, for instance, the long axis of the thigh is a real and unchangeable axis.

10 Bibliography

Ayres AJ (1979) Lernstörungen. Springer, Berlin Heidelberg New York

Benninghoff A (1971) Lehrbuch der Anatomie des Menschen. Urban and Schwarzenberg, Munich

Chapchal G (1971) Grundriß der orthopädischen Krankenuntersuchung. Enke, Stuttgart

Cyriax J (1969) Textbook of orthopaedic medicine. Bailliere Tindall and Cassel, London

Debrunner HU (1971) AO-Gelenkmessung. Neutral-0-Methode. Documentation of DGOT Bulletin, official organ of the Arbeitsgemeinschaft für Osteosynthesefragen, Tübingen

Frauenfelder P, Huber P, Staub H (1968) Einführung in die Physik, vol 1. Reinhardt, Basel

Hoeppke H (1971) Muskelspiel des Menschen. Fischer, Stuttgart

Kendall HD, Wadsworth GE (1971) Muscle testing and function. William and Wilkins, Baltimore

Klein-Vogelbach S (1981) Ballgymnastik zur funktionellen Bewegungslehre. Springer, Berlin Heidelberg New York

Klein-Vogelbach S (1990) Gangschulung zur funktionellen Bewegungslehre. Springer, Berlin Heidelberg New York (in press)

Klein-Vogelbach S (1990) Therapeutic Exercises. Springer, Berlin Heidelberg New York (in press)

Knott M (1969) Proprioceptive neuromuscular facilitation. Hoeber-Harper, New York

Kollmann (1901) Plastische Anatomie für Künstler. Veit, Leipzig

Lang J, Wachsmuth W (1972) Bein und Statik. Springer, Berlin Heidelberg New York (Praktische Anatomie, vol 1, part 4)

Lanz T von, Wachsmuth W (1959) Arm. Springer, Berlin Heidelberg New York (Praktische Anatomie, vol 1, part 3)

Mantler JT, Gatz AJ (1958) Clinical neuroanatomy and neurophysiology. Davis, Philadelphia

Müller M (1957) Die hüftnahen Femurosteotomien. Thieme, Stuttgart

Mumenthaler M (1965) Läsionen peripherer Nerven. Thieme, Stuttgart

Mumenthaler M (1980) Der Schulter-Arm-Schmerz. Huber, Bern

Pohl RW (1969, 1975, 1976) Einführung in die Physik, 3 vols. Springer, Berlin Heidelberg New York

Rickenbacher J, Landold AM, Theiler K (1985) Applied anatomy of the back. Springer, Berlin Heidelberg New York

Roederer JG (1977) Physikalische und psychoakustische Grundlagen der Musik. Springer, Berlin Heidelberg New York

Tittel K (1974) Beschreibende und funktionelle Anatomie. Fischer, Jena

Toldt-Hochstetter F (1975) Anatomischer Atlas. Urban and Schwarzenberg, Munich

11 Subject Index

constitution, length   216
-, width   217
contracture, muscular   54
counteractivity   110, 117
counter-rotational   42
counterweight   110, 117
cramp   287
cranioduction of shoulder joint   197
cranial (cranially)   19
critical distance point *see* point
- fulcrum   111
crus recurvatum   248, 269
crutch   98, 99

degree of freedom   37
depth   218
detorsion   244
device, supportive   7, 140, 149
-, suspension   7, 102, 103, 140, 149
diagonals, body   24
diameter, frontotransverse   28
-, sagittotransverse   28
-, thorax   28, 30
distal   34
distance point *see* point
distraction type   312
dorsal (dorsally)   21
dorsiduction of the shoulder joint   200
dorsiflexion   190

economical activity   81
- - intensity   83, 84
elbow joint *see* joint
end-stop   148, 312
equilibrium reaction   109
-, sophisticated   123
-, spontaneous   123
eversion   191
exercise, conception   147
extension   202, 226
- syndrome   253

fixed points, space   152
flexion   202, 225
-, lateral   228
foot, arch   248
-, functional long axis   176
-, longitudinal arch   176, 177, 245
-, transverse arch   177
forearm   206
-, pronation   207
free play   99, 100, 104, 107
- - function   100
frictional resistance   322
functional analysis   108
- problem   213, 286, 289
- status   213

- -, constitution mobility   213
- -, postural statics   213

gait test   265, 289
- -, basic   289
genu recurvatum   248, 269

hand, test position   212
-, - -, claw hand   212
-, - -, clenched fist   212
-, - -, fine-movement position   212
-, - -, functional clenched fist   212
-, - -, - opening   212
-, - -, opening the fist   212
hanging activity   101
hinge type   38, 63, 311
hip joint *see* joint
homunculus, the man in a cube   17
horizontal component   117
hypertonus, reactive   285, 287
hypothetical norm   140, 213
- -, deviation   213

instruction   144
-, manipulative   154
-, perceptual   154, 154
-, verbal   154
inversion   191
ischaemia   287
ischaemic pain   247, 254, 287
ischiocrural shortening   242
isometric training   127

joint
-, acromioclavicular   194, 253
-, atlanto-axial   253
-, - occipital   253
-, carpometacarpal (CM)   210
-, chopart   192
-, distal interphalangeal (DIP)   210
-, elbow   206
-, fingers   210
-, hip   234, 238
-, humeroscapular   196, 253
-, interphalangeal   212
-, knee   187, 248
-, -, valgus deviation   267
-, -, varus deviation   267
-, levels of movement   35
-, lisfranc   192
-, metacarpophalangeal (MP)   194, 210, 212
-, phalangeal   210
-, pivots of movement   35
-, proximal interphalangeal (PIP)   210
-, shoulder, abduction syndrome   222
-, -, caudalduction   198
-, -, cranialduction   197

tentacles 78
therapist language 144
- -, functional analysis 147
therapy 289
tibial torsion 271
track width 221, 277
translation 233
- type 312
translatory joint, compressive components 66
- -, motive components 66
- type 64
transverse extension 204
- flexion 203
treatment, plan 286
-, techniques 301

ulnar styloid process 194

valgus, femoral neck 269
- line 266
varus, femoral neck 269
- line 266
varus/valgus position, knee joint 258
ventral (ventrally) 21
ventroduction of the shoulder joint 199
vertebral column 251

weight, distribution 246
weight-bearing phase 278

P. M. Davies, Bad Ragaz

Steps To Follow

A Guide to the Treatment
of Adult Hemiplegia

Based on the Concept of K. and B. Bobath

With a Foreword by W. M. Zinn

1984. XXIII, 300 pp. 326 figs. in 492 sep. illus.
Softcover ISBN 3-540-13436-0

"...The author who shares her years of experience
with the reader is to be complimented on the
clarity of the text as well as the content. This is an
extremely easy book to use and must surely
become a standard text for all therapists working
with stroke patients wishing to use the Bobath
philosophy.
It is greatly helped by the use of excellent photographs which leave the reader in no doubt as to
the advice in the text.
This book should find a place in all physiotherapy
departmental libraries as well as schools of physiotherapy and no doubt physiotherapists working
with the stroke patient will want their own
personal copy."
Physiotherapy Practice

Also available in German:

Hemiplegie

Springer-Verlag Berlin
Heidelberg New York London
Paris Tokyo Hong Kong

1986. (Rehabilitation und Prävention, Band 18)
ISBN 3-540-12230-3

P. M. Davies

Right in the Middle

Selective Trunk Activity

1989. Approx. 250 pages. Approx. 500 illus.
In preparation
ISBN 3-540-51242-X

Every physiotherapist already acquainted with Pat Davies' best-selling guide to the treatment of adult hemiplegia, **Steps to Follow,** will want to own this book. It focuses on a subject that has been all but ignored up to now in rehabilitation literature: selective trunk activity.
The author once again shares her vast experience in treating hemiplegic patients and points to the amazing results that can be achieved when specific therapy to retrain and regain selective trunk activity is integrated into the treatment program. The key to successful treatment lies in the regaining of adaptive stabilisation of the trunk, and the ability to move parts of it in isolation.
This book gives clear, concise information to help physiotherapists learn how to observe and treat selective trunk activity.

Springer-Verlag Berlin
Heidelberg New York London
Paris Tokyo Hong Kong

Springer